ON THE
JOB
SERIES

REAL PEOPLE WORKING *in*

BUILDING AND CONSTRUCTION

Blythe Camenson

VGM Career Horizons
NTC/Contemporary Publishing Group

Library of Congress Cataloging-in-Publication Data

Camenson, Blythe.
    On the job : real people working in building and construction /
Blythe Camenson.
        p.   cm. — (On the job series)
    ISBN 0-8442-1927-4 (cloth). — ISBN 0-8442-1932-0 (pbk.)
    1. Construction industry—Vocational guidance. 2. Building
trades—Vocational guidance. I. Title. II. Series.
TH159.C36 1999
690′.023—dc21   98-46769
                                        CIP

To Pete Ziegler, who helped me build a business
as well as a friendship

Published by VGM Career Horizons
A division of NTC/Contemporary Publishing Group, Inc.
4255 West Touhy Avenue, Lincolnwood (Chicago), Illinois 60646-1975 U.S.A.
Copyright © 1999 by NTC/Contemporary Publishing Group, Inc.
Printed in the United States of America
International Standard Book Number: 0-8442-1927-4 (cloth)
                                      0-8442-1932-0 (paper)
99 00 01 02 03 04 VL 18 17 16 15 14 13 12 11 10 9 8 7 6 5 4 3 2 1

Contents

Acknowledgments

The author thanks the following professionals for providing information about their careers:

- Joel André, architect

- Peter Benton, architect

- Dan Brawner, painter

- Dick Chance, construction inspector

- Joseph Elliott, steam and sprinkler fitter

- Gene Formy-Duval, licensed general contractor

- Joy Formy-Duval, project scheduler

- Michael Keith, paperhanger

- Max King Jr., master plumber

- Kris Irr, roofing contractor

- Steve Lazarian Jr., licensed general contractor

- David McCafferty, brick mason

- Troy Scott McClure, refrigeration, heating, and air-conditioning contractor

- Holly Perry, drywall hanger

- Darrell Phelps, construction inspector

- Lawney Schultz, carpenter

- Lee Sullivan Hill, cost estimator

- Matthew I. Zehnder, landscape architect

How to Use This Book

On the Job: Real People Working in Building and Construction is part of a series of career books designed to help you find essential information quickly and easily. Unlike other career resources on the market, books in the *On the Job* series provide you with information on hundreds of careers, in an easy-to-use format. This includes information on

- The nature of the work

- Working conditions

- Employment

- Training, other qualifications, and advancement

- Job outlooks

- Earnings

- Related occupations

- Sources of additional information

But that's not all. You'll also benefit from a firsthand look at what jobs are really like, as told in the words of the employees themselves. Throughout the book, one-on-one interviews with practicing professionals enrich the text and enhance your understanding of life on the job.

These interviews tell what each job entails, what the duties are, what the lifestyle is like, and what the upsides and downsides are. All of the professionals reveal what drew them to the field and how they got started. And to help you make the best career choice for yourself, each professional offers you some expert advice based on years of experience.

Each chapter also lets you see at a glance, with easy-to-use reference symbols, the level of education required and salary range for the featured occupations.

So, how do you use this book? Easy. You don't need to run to the library and bury yourself in cumbersome documents from the Bureau of Labor Statistics, nor do you need to rush out and buy a lot of bulky books you'll never plow through. All you have to do is glance through our extensive table of contents, find the fields that interest you, and read what the experts have to say.

Introduction to the Field

Chances are, no matter where you live in North America, the sounds of a new construction site aren't too far away. As a people, we are builders, starting back with our own nation's history several hundred years ago.

If building and construction is in your blood, then you have probably already considered a career in one of the many areas of this wide-open field. Glancing through the table of contents will give you an idea of all the choices open to you.

But perhaps you're not sure of the working conditions the different trades offer or which area would suit your personality, skills, and lifestyle the most. There are several factors to consider when deciding which sector of a building and construction career to pursue. Each area carries with it different levels of responsibility and commitment. To identify occupations that will match your expectations, you need to know what each job entails.

Ask yourself the following questions and make note of your answers. Then, as you go through the chapters, compare your requirements with the information provided by the professionals interviewed for the various sectors. Their comments will help you pinpoint the fields that would interest you and eliminate those that would clearly be the wrong choice.

- How much of a people person are you? Do you prefer to work face-to-face with clients or customers, or are you more comfortable working just with your hands?

- Do you want a desk job, or would you prefer to be in the out-of-doors? Some occupations offer more freedom of movement than others.

- How much time are you willing to commit to training? Some skills can be learned on the job or in a year or two of formal training; others can take longer.

- How much money do you expect to earn, starting out and after you have a few years' experience under your belt? Salaries and earnings vary greatly in each trade and in different parts of the country. Some entry-level jobs pay very little; others will earn you a much higher salary.

- How much independence do you require? Do you want to be your own boss, or will you be content as a salaried employee?

- Will you work normal hours? Or will your day start at 5:30 A.M. and not end until 13 or 14 hours later? In a crunch can you handle working on holidays and weekends, or do you want that time free?

- What about working flat out for three to six months of the year, and then not knowing where the next paycheck will come from during the remaining months?

- How much physical labor can you handle? Would you prefer to avoid work that could be hard on your body?

Knowing what your expectations are, then comparing them with the realities of the work, will help you make informed choices.

Although *On the Job: Real People Working in Building and Construction* strives to be as comprehensive as possible, not all jobs in this extensive field have been covered or given the same amount of emphasis. You can find out more information by contacting the sources listed at the end of each chapter. You can also find professionals on your own to talk to and observe as they go about their work. Any remaining gaps you discover can be filled by referring to the *Occupational Outlook Handbook,* available on-line or at public libraries.

CHAPTER 1 Architects

EDUCATION
B.A./B.S.

$$$ SALARY
$26,000 to $110,000

OVERVIEW

Architects design buildings and other structures. The design of a building involves far more than its appearance. Buildings must also be functional, safe, and economical and must suit the needs of the people who use them. Architects take all these things into consideration when they design buildings and other structures.

Architects provide a wide variety of professional services to individuals and organizations planning a construction project. They may be involved in all phases of development, from the initial discussion of general ideas with the client through construction. Their duties require a number of skills: design, engineering, managerial, communication, and supervisory.

The architect and client first discuss the purposes, requirements, and budget of a project. Based on the discussions, architects may prepare a program—a report specifying the requirements the design must meet. In some cases, the architect assists in conducting feasibility and environmental impact analyses and selecting a site. The architect then prepares drawings and written information presenting ideas for the client to review.

After the initial proposals are discussed and accepted, architects develop final construction plans. These plans show the building's appearance and details for its construction. Accompanying these are drawings of the structural system; air-conditioning, heating, and ventilating systems; electrical systems; plumbing; and possibly site and landscape plans. They also specify the building materials and, in some cases, the interior furnishings. In developing designs, architects follow building codes, zoning laws, fire regulations, and other ordinances, such as those that require easy access by disabled persons. Throughout the planning stage, they make necessary changes. Although they have traditionally used pencil and paper to produce design and construction drawings, architects are increasingly turning to computer-aided design and drafting (CADD) technology for these important tasks.

Architects may also assist the client in obtaining construction bids, selecting a contractor, and negotiating the construction contract. As construction proceeds, they may be employed by the client to visit the building site to ensure that the contractor is following the design, meeting the schedule, using the specified materials, and meeting the specified standards for the quality of work. The job is not complete until all construction is finished, required tests are made, and construction costs are paid.

Architects design a wide variety of buildings, such as office and apartment buildings, schools, churches, factories, hospitals, houses, and airport terminals. They also design multibuilding complexes such as urban centers, college campuses, industrial parks, and entire communities. In addition to designing buildings, they may advise on the selection of building sites, prepare cost analysis and land-use studies, and do long-range planning for land development.

Architects sometimes specialize in one phase of work. Some specialize in the design of one type of building—for example, hospitals, schools, or housing. Others specialize in construction management or the management of their firm and do little design work. They often work with engineers, urban planners, interior designers, landscape architects, and others.

During a training period leading up to licensure as architects, entry-level workers are called intern-architects. This training period gives them practical work experience while they prepare for the Architect Registration Examination. Typical duties include preparing construction drawings on CADD, assisting in the design of one part of a project, and managing the production of a small project.

Architects generally work in a comfortable environment. Most of their time is spent in offices advising clients, developing reports and drawings, and working with other architects and engineers. However, they also often work at construction sites to review the progress of projects.

Architects may occasionally be under great stress, working nights and weekends to meet deadlines; a forty-hour workweek, however, is usual.

TRAINING

All states and the District of Columbia require individuals to be licensed (registered) before they may call themselves architects or contract to provide architectural services. Many architecture school graduates work in the field even though they are not licensed. However, a licensed architect is required to take legal responsibility for all work.

Three requirements generally must be met for licensure: a professional degree in architecture, a period of practical training or internship (usually for three years), and passage of all sections of the Architect Registration Examination.

In many states, the professional degree in architecture must be from one of the approximately one hundred schools of architecture with programs accredited by the National Architectural Accrediting Board (NAAB). However, state architectural registration boards set their own standards, so graduation from a nonaccredited program may meet the education requirement for licensure in some states. There are several types of professional degrees in architecture. The majority of all architecture degrees are from five-year Bachelor of Architecture programs intended for students entering from high school or with no previous architecture training. Some schools offer a two-year Master of Architecture program for students with a preprofessional undergraduate degree in architecture or a related area, or a three- or four-year Master of Architecture program for students with a degree in another discipline. In addition, there are many combinations and variations of these degree programs.

The choice of degree type depends on each individual's preference and educational background. Prospective architecture students should carefully consider the available options before committing to a program. For example, although the five-year Bachelor of Architecture program offers the fastest route to the professional degree, courses are specialized, and if the student does not complete the program, moving to a nonarchitecture program may be difficult. A typical program comprises courses in architectural history and theory; building design, including its technical and legal aspects; professional practice; math; physical sciences; and liberal arts. Many architecture schools also offer graduate education for those who already have a bachelor's or master's degree in architecture or other areas. Although graduate education beyond the professional degree is not essential for practicing architects, it is normally required for research, teaching, and certain specialties.

Architects must be able to visually communicate their ideas to clients. Artistic and drawing ability is very helpful in doing this, but not essential. More important is a visual orientation and the ability to conceptualize and understand spatial relationships. Good communication skills (both written and oral), the ability to work independently or as part of a team, and creativity are important qualities for anyone interested in becoming an architect. Computer literacy is also required, as most firms use computers for word processing, writing specifications, two- and three-dimensional drafting, and financial management. A knowledge of computer-aided design and drafting (CADD) is helpful and will become more important as architecture firms continue to adopt this technology.

New graduates usually begin in architecture firms, where they assist in preparing architectural documents or drawings. They also may do research on building codes and materials; or write specifications for building materials, installation criteria, the quality of finishes, and other related details. Graduates with degrees in architecture also enter related fields such as graphic, interior, or industrial design; urban planning; real estate development; civil engineering; or construction management.

In large firms, architects may advance to supervisory or managerial positions. Some architects become partners in established firms; others set up their own practices.

JOB OUTLOOK

Architects hold about 91,000 jobs. Most jobs are in architecture firms—the majority of which employ fewer than five workers.

Nearly one-third of architects are self-employed, practicing as partners in architecture firms or on their own. A few work for builders, real estate developers, and government agencies responsible for housing, planning, or community development, such as the U.S. Departments of Defense, Interior, and Housing and Urban Development, and the General Services Administration.

Architects' employment has traditionally been affected by the level of local construction, particularly of noninstitutional structures such as office buildings, shopping centers, schools, and health-care facilities. The boom in construction of commercial office space and some other types of nonresidential structures during the 1980s means there will be less construction of this type between now and 2005. Nevertheless, employment growth of architects is expected to increase as fast as the average for all occupations during this period.

The needed renovation and rehabilitation of old buildings, particularly in urban areas where space for new buildings is becoming limited, is expected to provide jobs for architects and to compensate somewhat for any slowdowns in jobs related to new construction. Also, the expected expansion of the population under age 15 and over age 65 should spur the demand for public and private buildings, such as schools and health-care facilities.

The need to replace architects who retire or leave the labor force for other reasons will provide many additional job openings.

Despite expected employment growth and the increased number of openings due to replacement needs, prospective architects may face competition, especially if the number of architecture degrees awarded remains at, or exceeds, current levels.

Traditionally, many individuals are attracted to this occupation, and there are often numerous applicants for available openings, especially in the most prestigious firms. Because non-institutional construction is sensitive to cyclical changes in the economy, architects will face particularly strong competition for jobs or clients during recessions, and layoffs may occur.

Those involved in the design of institutional buildings such as schools, hospitals, nursing homes, and correctional facilities will be less affected by fluctuations in the economy.

Even in times of overall good opportunities, there may be areas of the country with poor opportunities. Architects who are licensed to practice in one state must meet the licensing requirements of other states before practicing elsewhere. These requirements are becoming more standardized, however, facilitating movement to other areas of the country.

Because the use of computer-aided design and drafting is becoming more prevalent in architecture firms, prospective architects who know CADD technology may experience better opportunities in the future, particularly in a competitive job market.

SALARIES

According to the American Institute of Architects, the median salary for intern-architects in architecture firms is about $26,000. Licensed architects with 8 to 10 years' experience but who are not managers or principals of a firm earn a median salary of about $40,000, and principals or partners of firms earn a median salary of about $52,000. Partners in some large practices earn more than $110,000.

Most employers of wage and salary architects offer paid vacation and sick leave, and a majority also provide medical and life insurance plans to their employees. Employees of very small architecture firms (fewer than five employees) are less likely to receive these benefits.

Architects who are partners in well-established architecture firms generally earn much more than their salaried employees, but their income may fluctuate because of changing business conditions. Some architects may have difficulty getting established in their own practices and may go through a period in which their expenses are greater than their income, requiring substantial financial resources.

RELATED FIELDS

Architects design and construct buildings and related structures. Others who engage in similar work are landscape architects, building contractors, civil engineers, urban planners, interior designers, industrial designers, and graphic designers.

INTERVIEW
Joel André
Architect

Joel André is an architect and interior designer and owns his own firm called André Marquez Architects, located in Virginia Beach, Virginia. He has been practicing architecture since 1979, when he graduated from the School of Architecture at Pratt Institute in Brooklyn, New York. He has a Bachelor of Architecture degree, which is a professional degree requiring five years of schooling. In 1988 he passed the New York state licensing boards and became a licensed architect in the state of New York, and through reciprocity, he acquired his Virginia license, allowing him to practice in Virginia.

How Joel André Got Started

"I became interested in this profession after being exposed to it by an uncle who was an architect. So, from a very young age, I can remember wanting to do what he did. As it turns out, among all the cousins and distant cousins, there are six other architects and a number of engineers in my family.

"I have always been fascinated with the ability to design a structure that will meet a specific human need. I have always enjoyed the problem-solving aspect of the profession.

"I got my first job in the profession when I came out of school, working for a local architect in New York who had also graduated from Pratt. It was a small firm specializing in renovation and use conversion in New York, but primarily in Park Slope, Brooklyn, which is a brownstone type of neighborhood that was in the process of becoming 'gentrified.'

"The wonderful thing about working in this firm was that because it was so small, from the very beginning, I was able to get involved with the actual design of projects and be exposed to every facet of the profession.

"Some of the people I graduated with followed a different track. They went to large firms, where they were exposed to larger or more prominent projects but did not get, in my opinion, the full exposure to the nuts and bolts of the business that I got, as fast as I got it. In the larger firms your growth is a bit slower, but the advantage is that you get to work on some major projects. There are pros and cons in either track.

"Four years ago, I was in transition and out of a position as a project architect in a firm in Virginia Beach, when I decided to take the plunge and start my own firm. Because of the relationships that I had in the profession and past clients that I had remained in contact with, I was able to open my doors and hang my shingle without too much struggle. Because the firm I was working for closed its doors, there was no conflict in servicing some of their old clients. Like all new businesses, we go through some tight times, but on the average, we are doing better than would be expected."

What the Job's Really Like

"Our clients include commercial and residential accounts, churches, the Department of Defense, as well as state and municipal organizations. Because of the diversity of our clients, the types of projects that we are involved with also tend to be varied. That makes the day-to-day aspect of the business very interesting. But the bottom line of this profession is that it is a service-oriented field. It requires a desire to serve people, because ultimately, most buildings that you will be involved in designing will be used by people to facilitate or meet a specific need.

"To give you a broad-brush look at the practice of architecture, let me first run through the process for a typical project:

"One of the indispensable facets of the business of architecture is marketing. This involves following leads on who is planning to build or, even earlier, who is thinking of building. If you do work for the state or federal government, this involves responding to Requests for Proposals (RFPs) that are published in special publications or advertised in the local papers.

"Once you respond to the advertisement requesting architectural services, a selection process occurs, since there are generally several architectural firms pursuing these ads.

"Assuming that you are selected, a contract is negotiated with the client, establishing the scope of the project and your fee for the services that you will provide. The contract also establishes the schedule in which the construction documents will be accomplished. These documents include the drawings, specifications, and coordination of the other disciplines that are going to be involved in the project. Generally these are other engineering fields, such as civil engineers for the site work, structural engineers for the foundations and structure framework of the building, the plumbing and mechanical, meaning heating and air-conditioning, and electrical engineers, and also the landscape architect, the fire protection engineer for sprinkler systems, and any other special conditions that require input from a specialist, such as acoustical conditions or lighting conditions.

"The architect is responsible for the production of the documents that will be used by the builder or contractor to build the building. He or she is responsible for making sure that the program, or the list of requirements and needs and functions that the building has to meet, satisfies all the needs of the owner/user, from a functional point of view as well as in terms of the safety and welfare of the occupants.

"The architect also must do all of this within budget—which often requires a reevaluation of the scope of the project to fit the budget or an adjustment of the budget to fit the scope. It is critical that this occurs in the beginning stages of the project.

You may design a great building, but if it can't be built because it exceeds the budget, it is a bad experience for everyone. So, a very important facet of this business is cost estimating.

"Once the scope of the project and the budget for the project have been reconciled and the architect has pulled together a team of consultants, we will launch into the concept design phase of the project. This is where the preliminary design work is done. We establish the general direction of the shape, volume, basic space layouts, and site layout, showing where on the site the building sits. We take into consideration access, visibility, presentation to street, relationship to neighbors, solar orientation, drainage of site, utility connections, and what the building will look like, with consideration for materials to be used, while complying with all the zoning and building codes applicable.

"During this phase and all the other phases of this process, we do a constant review of the cost and budget to make sure that the decisions being made in the design process are going to be buildable.

"Once the concept design has been reviewed and approved by the client (this generally requires several presentations and revisions), the project goes into the design development phase. This is where the design is refined, materials are selected, the engineer consultants are given directions to do their portion of the work, and the coordination of these various disciplines begins in earnest.

"At the end of this phase, the building design is finalized and, as always, the evaluation of the cost versus budget is done.

"From this phase, we proceed to the construction document production phase. This is where the construction drawings or blueprints are done, as well as the specifications that spell out in more detail the materials, the quality of the materials, the installation of these materials, and the maintenance of the materials, as well as the warranties for each material to be used in the building.

"The specifications complement the drawings in order to give the contractor all the information needed to build the building as it is designed. This phase generally is the longest.

"While the architect is producing the drawings and specifications, the consultants are also producing their drawings and documents that will complement the architectural documents. These are added to the architectural drawing set, and their respective specification sections are added to the project specification book.

"Once the construction documents are completed, they are released so that contractors interested in building the project can make their bids.

"Once a contractor has been selected, the responsibility of the architect is to make sure that the client is getting what he or she is paying for and that the contractor is using the materials and installing the equipment that were specified in the construction documents. At the same time, the architect has a responsibility to make sure that the contractor has all the information needed to do the work.

"Once the building is completed and all systems are operational, the project is closed out, and the owner/user assumes occupancy of the building. Ideally, all parties are satisfied, and we all part friends. This, I find, depends greatly on the individual players in this complex process. However, we do carry liability insurance.

"Generally, there are no typical days. Depending on which phase of a project you are in, you have specific goals you are working toward. And it gets even more interesting when you have several projects in various phases of development. This is where a good staff of designers, CAD operators, administrative support, project architects, and project managers come into play.

"Also, depending on the types of projects that you do, the approach to the work might vary. Some firms specialize in one building type, such as hospitals or hotels or residences.

"However, on the average, because of my position, I spend a lot of time coordinating the work that goes on in the office among the client, consultants, contractors, and all other interested parties. I also do the concept design for the projects. I want that initial input, establishing the direction of the design to the satisfaction of the client.

"This is a people profession. If my client isn't happy, I am not going to be happy, since the client will find himself or herself another architect.

"What I like most about practicing architecture is the creative aspect of the business. The design and problem-solving aspects are the things that I like to do. I love to design and conceptualize a client's vision into a product that fulfills that vision. I enjoy working out the details in the design development phase. Unfortunately, that's only about 10 to 15 percent of the whole process.

"I also enjoy the construction document phase of the work, the working drawings and the specifications, but not as much as the design.

"The part that I enjoy the least is the administrative or business end of the profession. That is not why I became an architect, but it is an unavoidable part of the business. In fact, I would suggest that an M.B.A. would be an ideal postgraduate degree for architects. Either that or hire a good business manager. We have to wear so many hats as architects that trying to wing it as a businessperson can be deadly.

"The upside of architecture is that it is a profession that will give you a view on your environment that is very different from most other professions. You will have an understanding of your environment that no other profession will give you.

"I also find that it affects every aspect of your life, because remaining in the profession does require a love for it.

"The financial rewards are not always there, especially when you're starting out. Most architects are not rich. I earn approximately $50,000 a year. My salary is based on the fees I negotiate with the client at the onset of the contract.

"An individual just out of school can expect to make anywhere from $22,000 to $28,000 per year. This varies depending on the geographic location and the state of the economy. Large cities tend to have higher salaries, but the cost of living is generally higher, so it's often a wash.

"A project manager, or project architect, who is licensed can make approximately $32,000 to $50,000, again depending on the firm and the location. An established architect can earn anywhere from $50,000 to $150,000 a year, and in the larger firms with profit sharing and other benefit packages, the earnings can be even higher."

Expert Advice

"The primary requisites to be an architect are a good understanding of the profession and a love for the work. I say that because it does require a lot from you, often with small rewards, especially in the beginning.

"The qualities you'll need to be a good architect are good people skills, problem-solving skills, creativity, and an appreciation of the pure logic of engineering.

"For training, I would recommend a five-year school providing a professional degree, and practical office work during summer and winter breaks. Find an architect who will bring you in and maintain that relationship throughout your schooling. It's an incredible asset to have the right expectations of the profession when you come out of school, as opposed to stepping into the reality of everyday practice while your head is still in the clouds of academia.

"So, to get started, I recommend you visit an architect's office or, better yet, several offices, and if at all possible, spend a couple of weeks of a summer vacation in an architect's office, even if you have to volunteer your time. Then, I would visit some architectural schools to get a good understanding of the profession.

"Ultimately, make sure that it's something that you really want to do, and understand the sacrifices that it will require."

INTERVIEW
Peter Benton
Architect

Peter Benton is a restoration architect specializing in historic preservation. He earned a B.S. in architecture in 1972 from the University of Virginia in Charlottesville. He worked for several years for various firms in Philadelphia and Washington, D.C., then went on to complete his M. Arch. (Master of Architecture) from the University of Pennsylvania in 1979.

How Peter Benton Got Started

"Initially I had relatively little training in preservation, but I was exposed to the idea of ecological planning at University of Pennsylvania. I saw the philosophical connection between an ecological approach to the landscape and to the buildings, and that led me to historic preservation. I went to work for four or five years for an ecological planning firm, and it was there my interest developed further.

"I am now a senior associate with John Milner and Associates, Inc., a midsize architectural firm in West Chester, Pennsylvania, specializing in historic preservation. I joined the staff in 1984 and have worked on a variety of projects."

What the Job's Really Like

"The difference between a general architect and a restoration architect is that the latter's work experience has primarily been focused on historic buildings. In addition, the restoration architect will have a specialized knowledge and understanding of federal, state, and local regulations with regard to historic preservation. Restoration architects will also be aware of the standards set by the particular style of architecture.

"I've been responsible for all sorts of properties—anything from small, privately owned residential-scale houses from the 18th century to high-style 19th-century mansions. In addition, I've worked with historic commercial and industrial buildings from the 19th century, restoring them or practicing what we call adaptive reuse. For example, we recently converted an old mill into an office building and a farmhouse into a meeting facility. Another category I've worked with includes monumental buildings, such as a city hall or large federal buildings.

"First I meet with the client and determine what his or her goals for the property are. Then I do an existing-conditions analysis of the site, look at the historical development of the building over time, and take photographs, field measurements, and written notes.

"Next I do a schematic plan, making preliminary drawings and sketches, describing a design to the client for the client's approval. This stage could take four weeks or so. Once the client approves the project, I produce an outline of the scope of work and figure an order-of-magnitude cost estimate.

"After that stage is approved, the next phase is to work on design development documents. This involves the use of more detailed drawings and can take from six to eight weeks. Over the next eight to twelve weeks, construction documents including drawings and specifications are produced. During the bidding phase, the contractor is selected. Then if necessary, we review and revise the construction plans before actually beginning work. I make frequent visits to the site while the project is in progress.

"Construction time varies but could take eight months or a year and a half depending upon the scope of the project."

Expert Advice

"Someone fresh from graduate school can expect to earn from $25,000 to $27,000 per year, depending of course on the size of the firm, the importance of the project, and the region of the country. Advancement would depend on your ability and accomplishments.

"An experienced architect with five years or more at the project manager level could expect to earn about $40,000 a year in a midsize firm. Those with specializations that are in demand can earn more. Most firms offer paid internships for graduate students, and it is a good idea to set one up for yourself. It gives you a foot in the door when it comes time to make arrangements for a full-time job when you graduate."

FOR MORE INFORMATION

Information about education and careers in architecture can be obtained from:

Architecture Fact Book
American Institute of Architects
1735 New York Avenue NW
Washington, D.C. 20006

Society of American Registered Architects
1245 South Highland Avenue
Lombard, IL 60148

CHAPTER 2 Landscape Architects

EDUCATION
B.A./B.S. Required

SALARY
$40,000 to $50,000

OVERVIEW

Everyone enjoys attractively designed residential areas, public parks, college campuses, shopping centers, golf courses, parkways, and industrial parks. Landscape architects design these areas so that they are not only functional but beautiful and compatible with the natural environment as well. They may plan the location of buildings, roads, and walkways and the arrangement of flowers, shrubs, and trees. Historic preservation and natural resource conservation and reclamation are other important objectives to which landscape architects may apply their knowledge of the environment as well as their design and artistic talents.

Landscape architects are hired by many types of organizations—from real estate development firms starting new projects to municipalities constructing airports or parks. They are often involved with the development of a site from its conception.

Working with architects, engineers, scientists, and other professionals, they help determine the best arrangement of roads and buildings, and the best way to conserve or restore natural resources. Once these decisions are made, landscape architects create detailed plans indicating new topography, vegetation, walkways, and landscape amenities.

In planning a site, landscape architects first consider the nature and purpose of the project and the funds available. They analyze the natural elements of the site, such as the climate, soil, slope of the land, drainage, and vegetation. They observe where sunlight falls on the site at different times of the day and examine the site from various angles. They also assess the effect of existing buildings, roads, walkways, and utilities on the project.

After studying and analyzing the site, they prepare a preliminary design. To account for the needs of the client as well as the conditions at the site, they may have to make many changes before a final design is approved. They must also take into account any local, state, or federal regulations, such as those protecting wetlands or historic resources.

An increasing number of landscape architects are using computer-aided design (CAD) systems to assist them in preparing their designs. Many landscape architects also use video simulation as a tool to help clients envision the proposed ideas and plans. For larger-scale site planning, landscape architects also use geographic information systems technology, a computer mapping system.

Throughout all phases of the planning and design, landscape architects consult with other professionals involved in the project. Once the design is complete, they prepare a proposal for the client. They produce detailed plans of the site, including written reports, sketches, models, photographs, land-use studies, and cost estimates, and submit them for approval by the client and by regulatory agencies. If the plans are approved, landscape architects prepare working drawings showing all existing and proposed features. They also outline in detail the methods of construction and draw up a list of necessary materials.

Although many landscape architects supervise the installation of their designs, some are also involved in the construction of the site. However, this usually is the responsibility of the developer or landscape contractor.

Some landscape architects work on a wide variety of projects. Others specialize in a particular area, such as residential development, historic landscape restoration, waterfront improvement projects, parks and playgrounds, or shopping centers. Still others work in regional planning and resource management; feasibility, environmental impact, and cost studies; or site construction. Some landscape architects teach in colleges or universities.

Although most landscape architects do at least some residential work, relatively few limit their practice to landscape design for individual home owners. The reason is that most residential landscape design projects are too small to provide suitable income compared with larger commercial or multiunit residential projects.

Some nurseries offer residential landscape design services, but these services often are performed by lesser-qualified landscape designers or others with training and experience in related areas.

Landscape architects who work for government agencies do similar work at national parks, government buildings, and other government-owned facilities. In addition, they may prepare environmental impact statements and studies on environmental issues such as public land-use planning.

Landscape architects combine their knowledge of design, construction, plants, soils, and ecology to create their final designs.

Landscape architects spend most of their time in offices creating plans and designs, preparing models and cost estimates, doing research, or attending meetings. The remainder of their time is spent at the site. During the design and planning stage, landscape architects visit and analyze the site to verify that the design can be incorporated into the landscape. After the plans and specifications are completed, they may spend additional time at the site observing or supervising the construction. Those who work in large firms may spend considerably more time out of the office because of travel to sites outside the local area.

Salaried employees in both government and landscape architectural firms usually work regular hours, although they may work overtime to meet a project deadline. Hours of self-employed landscape architects may vary.

TRAINING

A bachelor's or master's degree in landscape architecture is usually necessary for entry into the profession. The bachelor's degree in landscape architecture takes four years and focuses on laws and/or plant materials indigenous to the state.

Because states' requirements for licensure are not uniform, landscape architects may not find it easy to transfer their registration to another state. However, those who meet the national standard of graduating from an accredited program, serving three years of internship under the supervision of a registered landscape architect, and passing the L.A.R.E. (Landscape Architect Registration Examination) can satisfy requirements in most states.

In the federal government, candidates for entry positions should have a bachelor's or master's degree in landscape architecture. The federal government does not require its landscape architects to be licensed.

People planning a career in landscape architecture should appreciate nature and enjoy working with their hands. Although creativity and artistic talent are also desirable qualities, they are not absolutely essential to success as a landscape architect. High school courses in mechanical or geometric drawing, art, botany, and mathematics are helpful. Good oral communication skills are important, because these workers must be able to convey their ideas to other professionals and clients and to make presentations before large groups.

Landscape architects do research and prepare reports and land-impact studies, so strong writing skills are valuable. A knowledge of computer applications of all kinds, including computer-aided design and drafting (CADD), is becoming increasingly necessary. Those interested in starting their own firms should be skilled in small business management.

In states where licensure is required, new hires are technically called intern landscape architects until they become licensed. Their duties vary depending on the type and size of employing firm. They may do project research or prepare base maps of the area to be landscaped, while some are allowed to participate in the actual design of a project. However, interns must perform all work under the supervision of a licensed landscape architect. Additionally, all drawings and specifications must be approved by the licensed landscape architect, who takes legal responsibility for the work. After gaining experience and becoming licensed, landscape architects usually can carry a design through all stages of development. After several years, they may become associates, and eventually they may become partners in a firm or open their own offices.

JOB OUTLOOK

Landscape architects hold about 14,000 jobs. Three-fifths work for firms that provide landscape architecture services. Most of the rest are employed by architectural firms. The federal government also employs these workers; most are found in the U.S. Departments of Agriculture, Defense, and Interior.

Approximately one out of every five landscape architects is self-employed.

Most employment for landscape architects is concentrated in urban and suburban areas in all parts of the country. Some landscape architects work in rural areas, particularly those in the federal government who plan and design parks and recreation areas.

Despite expected stronger employment growth and higher replacement needs due to retirements than in the past decade, new graduates can expect to face competition for jobs as landscape architects. The number of professional degrees awarded in landscape architecture has remained steady over the years, even during times of fluctuating demand because of economic conditions. If this trend continues, the number of openings in this small occupation will be too few to absorb all job seekers.

Traditionally, however, people with landscape architecture training qualify for jobs closely related to landscape architecture and may become construction or landscape supervisors, landscape designers, drafters, land or environmental planners, or landscape consultants.

Opportunities will be best for those who develop strong technical skills and a knowledge of environmental issues, codes, and regulations.

Employment of landscape architects is expected to increase about as fast as the average for all occupations through the year 2005. The level of new construction plays an important role in determining demand for landscape architects. Anticipated growth in construction is expected to increase demand for landscape architectural services over the long run. An increasing proportion of office and other commercial and industrial development will occur outside cities. These projects are typically located on larger sites with more surrounding land that needs to be designed, in contrast to urban development, which often includes little or no surrounding land. Also, as the cost of land increases, the importance of good site planning and landscape design increases. Because employment is linked to new construction, however, landscape architects may face layoffs and competition for jobs when real estate sales and construction slow down, such as during a recession.

Increased development of open space into recreation areas, wildlife refuges, and parks will also require the skills of landscape architects. Continued concern for the environment should stimulate employment growth because of the need to design development projects that best fit in with the surrounding environment.

In addition to the work related to new development and construction, landscape architects are expected to be involved in historic preservation; local, city, and regional planning; land reclamation; and refurbishment of existing sites.

The need to replace landscape architects who retire or leave the labor force for other reasons is expected to result in nearly as many openings as new openings due to job growth.

SALARIES

According to a recent American Society of Landscape Architects survey, the median salary for landscape architects in private practice is about $40,000; the median bonus, $4,000; and additional landscape architecture–related income, $5,000.

Those who work in the public sector earn higher salaries— a median of $42,400—but median bonus amount and outside landscape architecture–related income are lower than for private practitioners.

The average annual salary for all landscape architects in the federal government in nonsupervisory, supervisory, and managerial positions is $49,570.

Because many landscape architects work for small firms or are self-employed, benefits tend to be less generous than those of other workers with similar skills who work for large organizations. With the exception of those who are self-employed, however, most landscape architects receive health insurance, paid vacations, and sick leave.

RELATED FIELDS

Landscape architects use their knowledge of design, construction, and land-use planning to develop a landscape project. Others whose work requires similar skills are architects, interior designers, civil engineers, and urban and regional planners. Landscape architects also know how to grow and use plants in the landscape. Botanists, who study plants in general, and horticulturists, who study ornamental plants as well as fruit, vegetable, greenhouse, and nursery crops, do similar work.

INTERVIEW
Matthew I. Zehnder
Senior Vice President of Landscape Architecture and Community Planning

Matthew Zehnder works for META Associates, Inc., in Louisville, Kentucky, a health-care-oriented strategic planning and program management firm. The firm deals primarily with all developmental aspects of health-care-related projects, from the master planning of health-care assisted living communities to 40-bed private-care facilities.

He earned his B.S.L.A. (Bachelor of Science in Landscape Architecture) from the University of Kentucky in Lexington in 1988. In 1993 *he earned a master's degree in landscape architecture and regional planning from the University of Pennsylvania in Philadelphia.*

How Matthew I. Zehnder Got Started

"I was first attracted to the profession because I thought it would allow me to work outside. I enjoyed working with plants and wanted to learn designing using plants as a palette.

"I researched several firms in the community where I wanted to live. I learned as much about each firm as possible. I then proceeded to learn about the partners of the firms, so that when I would go on my interviews, I could speak intelligently about the firms' accomplishments.

"Later I was approached by a mid-level executive from the firm I am now employed with. The person who approached me asked if I would be interested in leading their planning department. I was, indeed, interested and began what was to be a six-month interview process.

"I was interviewed by all three partners separately for approximately two hours per interview. I was essentially allowed to espouse my views and ideas concerning design and how I market myself and the firms I have represented in the past.

"After my last interview, I was asked to lunch with all the partners. They asked me about my specific career goals and what I would like to accomplish in my design life.

"After that, I was asked to draft a business plan and forecast a figure of billable hours for myself. Following my producing this document, there was a final interview during which I presented my requirements for employment. Our negotiations following this meeting culminated in a written offer that I accepted with conditions."

What the Job's Really Like

"My job is the most wonderful job I could have imagined! I set my own hours—within reason. I answer to one partner and am responsible for myself. My job is a little unusual in that I have pursued the marketing and contractual administration path that it seems few LAs pursue.

"My typical day begins at 9:00 in the morning. I listen to my voice mail and promptly answer them all. I then write around 10 notes to different business associates and friends so that my network line remains open and up-to-date. The three most important words: *network, network, network*!

"If I am not working on a proposal, then I am most likely either completing a schematic design or making several appointments to visit with people I think my firm could provide a service to.

"I do not spend a lot of time drawing. When I first began my LA career, I ran blueprints and mostly did the grunt work—you know, low man on the pole. I have progressed to executive management because that was my wish. I continue to speak with associates of mine, former classmates, who are still drafting for a living. It's pretty much a profession where you can set your own height.

"My job keeps me moving, both locally and nationally, and it's very interesting work. I subscribe to a simple philosophy: I will continue to work and do a good job for my employer as long as I enjoy what I am doing. When I no longer enjoy my job, I begin looking for new opportunities.

"My workweek averages about 65 hours. I do not bring work home! My position is very corporate, yet I depend on many individuals to assist me in completing my job; that is, you must be willing to believe that there are others who can do just as good a job as you, or better. I travel to the job sites frequently. My firm competes nationally, so I log approximately 120,000 miles a year in the sky. The work atmosphere is very user friendly: everyone freely comments about your design ideas and makes suggestions, when warranted, on ways to improve a design.

"I enjoy the freedom my employer permits me. I enjoy the trust my employer puts in me, and I enjoy the varied design professionals I interact with. The least enjoyable part of my job is the paperwork and the complicated time sheet processes."

Expert Advice

"Know what you want to do in life—where you want to be. It is essential to write down a set of goals to reach in five years and continue to review your goals and adjust them as necessary.

"You must network with all types of people, and you must decide between the academic world and the professional world of this profession—they are two very different worlds.

"Pick the firm you want to work for, and pursue it as if it's the last place in the world to work. And the most important item: believe in what you want to do, and do it!"

FOR MORE INFORMATION

Additional information, including a list of colleges and universities offering accredited programs in landscape architecture, is available from:

American Society of Landscape Architects
Career Information
4401 Connecticut Avenue NW, Suite 500
Washington, D.C. 20008

General information on registration or licensing requirements is available from:

Council of Landscape Architectural Registration Boards
12700 Fair Lakes Circle, Suite 110
Fairfax, VA 22033

CHAPTER 3 Construction Managers

OVERVIEW

Construction managers assume a wide variety of responsibilities and positions within construction firms. They are known by a range of job titles that are often used interchangeably—for example, construction superintendent, general superintendent, project manager, general construction manager, or executive construction manager. Construction managers may be owners or salaried employees of a construction management or contracting firm, or individuals working under contract or as salaried employees for the owner, developer, contractor, or management firm overseeing the construction project. Here we'll use the title "construction manager" to encompass all supervisory-level salaried and self-employed construction managers who oversee construction supervisors and workers.

In the construction industry, managers and other professionals active in the industry—general managers, project engineers, cost estimators, and others—are increasingly referred to as *constructors*. This designation applies to a broad group of professionals in construction who, through education and experience, are capable of managing, coordinating, and supervising the construction process from conceptual development through final construction on a timely and economical basis.

Given designs for buildings, roads, bridges, or other projects, constructors oversee the organization, scheduling, and implementation of the project to execute those designs. They are responsible for coordinating and managing people, materials, and equipment; budgets, schedules, and contracts; and the safety of employees and the general public.

The designation "construction manager" is also used more narrowly within the construction industry to denote a firm, or an individual employed by the firm, involved in management oversight of a construction project. Under this narrower definition, construction managers generally act as agents or representatives of the owner or developer throughout the life of the project. Although they generally play no direct role in the actual construction of the building or other facility, they typically schedule and coordinate all design and construction processes. They develop and implement a management plan to complete the project according to the owner's goals that allows the design and construction processes to be carried out efficiently and effectively within budgetary and schedule constraints.

In this book, "construction manager" includes these workers as well as managers working directly for the contractors who actually perform the construction.

Generally, a contractor is the firm under contract to provide specialized construction services. On small projects such as remodeling a home, the construction contractor is usually a self-employed construction manager or skilled trades worker who directs and oversees employees. On larger projects, construction managers working for a general contractor have overall responsibility for completing the construction in accordance with the engineer's or architect's drawings and specifications and prevailing building codes. They arrange for subcontractors to perform specialized craft work or other specified construction work.

Large construction projects, such as an office building or industrial complex, are too complicated for one person to supervise. These projects are divided into many segments: site preparation, including land clearing and earth moving; sewage systems; landscaping and road construction; building construction, including excavation and laying foundations, erection of structural framework, floors, walls, and roofs; and building systems, including fire protection, electrical, plumbing, air-conditioning, and heating. Construction managers may work as part of a team or may be in charge of one or more of these activities. They may have several subordinates reporting to them, such as assistant project managers, superintendents, field engineers, or crew supervisors.

Construction managers plan, budget, and direct the construction project. They evaluate various construction methods and determine the most cost-effective plan and schedule. They determine the appropriate construction methods and schedule all required construction site activities into logical, specific steps, budgeting the time required to meet established deadlines. This may require sophisticated estimating and scheduling techniques, using computers with specialized software. Construction managers also determine the labor requirements and, in some cases, supervise or monitor the hiring and dismissal of workers.

Managers direct and monitor the progress of field or site construction activities, at times through other construction supervisors. This includes the delivery and use of materials, tools, and equipment; the quality of construction; worker productivity; and safety. They are responsible for obtaining all necessary permits and licenses and, depending on the contractual arrangements, direct or monitor compliance with building and safety codes and other regulations.

They regularly review engineering and architectural drawings and specifications to monitor progress and ensure compliance with plans and specifications. They track and control construction costs to avoid cost overruns. Based on direct observation and reports by subordinate supervisors, managers may prepare daily reports of progress and requirements for labor, material, and machinery and equipment at the construction site.

Construction managers meet regularly with owners, subcontractors, architects, and other design professionals to monitor and coordinate all phases of the construction project.

They work out of a main office from which the overall construction project is monitored or they work out of a field office at the construction site. Management decisions regarding daily construction activities are usually made at the job site.

Managers usually travel when the construction site is in another state or when they are responsible for activities at two or more sites. Management of construction projects overseas usually entails temporary residence in another country.

Construction managers must be "on call" to deal with delays, bad weather, or emergencies at the site. Most work more than a standard 40-hour week, since construction may proceed around-the-clock. This type of work schedule can go on for days, even weeks, to meet special project deadlines, especially if there have been unforeseen delays.

Although the work generally is not considered dangerous, construction managers must be careful while touring construction sites, more so when large machinery, heavy equipment, and vehicles are being operated. Managers must be able to establish priorities and assign duties. They need to observe job conditions and to be alert to changes and potential problems, particularly involving safety on the job site and adherence to regulations.

TRAINING

Anyone interested in becoming a construction manager needs a solid background in building science and management, as well as related work experience within the construction industry. They need to be able to understand contracts, plans, and specifications, and be knowledgeable about construction methods, materials, and regulations. Familiarity with computers and software programs for job costing, scheduling, and estimating is increasingly important.

Traditionally, people advanced to construction management positions after having substantial experience as construction craft workers—for example, as carpenters, masons, plumbers, or electricians—or after having worked as construction supervisors or as independent specialty contractors overseeing workers in one or more construction trades. However, more and more employers—particularly, large construction firms—seek to hire managers with industry work experience and formal postsecondary education in building science or construction management.

In 1998, more than 100 colleges and universities offered four-year degree programs in construction management or construction science. These programs include courses in project control and development, site planning, design, construction methods, construction materials, value analysis, cost estimating, scheduling, contract administration, accounting, business and financial management, building codes and standards, inspection procedures, engineering and architectural sciences, mathematics, statistics, and computer science.

Graduates from four-year degree programs usually are hired as assistants to project managers, field engineers, schedulers, or cost estimators. An increasing number of graduates in related fields—engineering or architecture, for example—also enter construction management, often after having had substantial experience on construction projects.

Around 30 colleges and universities also offer a master's degree program in construction management or construction science, and at least two offer a Ph.D. in the field. Master's degree recipients, especially those with work experience in construction, typically become construction managers in very large construction or construction management companies. Often, individuals who hold a bachelor's degree in an unrelated field seek a master's degree in order to work in the construction industry. Doctoral degree recipients generally become college professors or work in an area of research.

Many individuals also attend training and educational programs sponsored by industry associations, often in collaboration with postsecondary institutions. A number of two-year colleges throughout the country offer construction management or construction technology programs.

Construction managers should be adaptable and be able to work effectively in a fast-paced environment. They should be decisive and able to work well under pressure, particularly when faced with unexpected occurrences or delays. The ability to coordinate several major activities at once, while being able to analyze and resolve specific problems, is essential, as is the ability to understand engineering, architectural, and other construction drawings. Managers must be able to establish a good working relationship with many different people, including owners, other managers, design professionals, supervisors, and craft workers.

Advancement opportunities for construction managers vary depending on the size and type of company. Within large firms, managers may eventually become top-level managers or executives. Highly experienced individuals may become independent consultants; some serve as expert witnesses in court or as arbitrators in disputes. Those with the required capital may establish their own firms, offering construction management services or their own general contracting firms overseeing construction projects from start to finish.

JOB OUTLOOK

Construction managers held approximately 249,000 jobs in 1996. Around 40,000 were self-employed. More than 85 percent were employed in the construction industry, primarily by specialty trade contractors—for example, plumbing, heating and air-conditioning, and electrical contractors—and general building contractors. Many also worked as self-employed independent contractors in the specialty trades. Others were employed by engineering, architectural, surveying, and construction management services firms, as well as local governments, educational institutions, and real estate developers.

Employment of construction managers is expected to increase faster than the average for all occupations through the year 2006 as the level of construction activity and complexity of construction projects continues to rise. In addition, many job openings should result annually from the need to replace workers who transfer to other occupations or leave the labor force.

Employers prefer applicants with construction work experience who can combine a strong background in building technology with proven supervisory or managerial skills. Prospects in construction management, engineering and architectural services, and construction contracting firms should be particularly favorable for people with a bachelor's degree or higher in construction science, construction management, or construction engineering who have worked in construction.

Increased spending on the nation's infrastructure—highways, bridges, dams, water and sewage systems, and electric power generation and transmission facilities—will result in a greater demand for construction managers, as will the need to build more residential housing, commercial and office buildings, and factories. Continuing maintenance and repair of all kinds of existing structures will also contribute to the demand for these professionals.

The increasing complexity of construction projects also should lead to the creation of more manager jobs. Advances in building materials and construction methods and the growing number of multipurpose buildings, electronically operated "smart" buildings, and energy-efficient structures will require the expertise of more construction managers.

In addition, the proliferation of laws setting standards for buildings and construction materials, worker safety, energy efficiency, and environmental pollution have further complicated the construction process and should increase demand for managers. As project owners and construction companies strive to keep costs in line and reduce the causes of disputes and litigation, they will continue to depend on the services and expertise of highly effective managers.

Employment of construction managers is sensitive to the short-term nature of many construction projects and cyclical fluctuations in construction activity. However, during periods of diminished construction activity—when many construction workers are laid off—many construction managers remain employed to plan, schedule, or estimate costs of future construction projects, as well as to manage ongoing maintenance, repair, and renovation work.

SALARIES

Earnings of salaried construction managers and incomes of self-employed independent construction contractors vary depending on the size and nature of the construction project, its geographic location, and economic conditions. According to a 1997 salary survey by the National Association of Colleges and Employers, bachelor's degree holders with degrees in the field of construction management received offers averaging $28,060 a year. Those with degrees in the field of construction science received offers averaging $31,949 a year.

Based on limited information available, the average starting salary for experienced construction managers in 1996 ranged from $40,000 to $100,000 annually.

Many salaried construction managers receive benefits such as bonuses, use of company motor vehicles, paid vacations, and life and health insurance.

RELATED FIELDS

Construction managers participate in the conceptual development of a construction project and oversee its organization, scheduling, and implementation. Occupations that perform similar functions include architects, civil engineers, construction supervisors, cost engineers, cost estimators, developers, electrical engineers, industrial engineers, landscape architects, and mechanical engineers.

INTERVIEW
Steve Lazarian Jr.
Licensed General Contractor

Steve Lazarian Jr. is owner and president of CityWorks, Inc., located in Pasadena, California. In 1970 he earned his B.A. in economics and business from Westmont College in Montecito, California, and in 1973 his law degree from Cal Western School of Law in San Diego. He is also a licensed real estate broker and has been in the field since 1972.

How Steve Lazarian Jr. Got Started

"I grew up in a construction atmosphere. My father started an electrical construction business in 1948. I worked as a kid, sweeping the warehouse, and then later in the office in purchasing, estimating, and accounting.

"The business became very successful, but I was not interested in the technical aspects of the business, so I went to law school. After graduating, passing the bar, and starting my own practice, I realized that I enjoyed construction law more than anything. I developed a clientele with contractors, subcontractors, developers, and real estate brokers.

"After my appointment to the Contractor's State License Board in 1985, I became interested in the business itself. Our family started a general construction business in 1983 that was becoming successful and growing fast. In 1989 I decided to leave my practice and manage the general construction company. That company was called Crown City Construction. It was liquidated in 1995. That same year, I established my own construction company called CityWorks.

"My training came from my business degree, law degree, practice of law, serving on the Contractor's Board, and working in my father's business."

What the Job's Really Like

"My duties include overseeing the jobs that we construct and the management of construction projects for which we're performing services before construction actually commences. I usually start my day by outlining my priorities of what I need to accomplish on that day. This includes confirming meetings and responsibilities of other members of the staff. I spend much of my time preparing spreadsheets, writing contracts, communicating with subcontractors, authorizing payments to subs and vendors, visiting job sites, and working with owners, architects, engineers, and related parties.

"Construction is a very high-risk business. You cannot afford to relax at any time. There is no time to even be 'busy-relaxed.' I need every moment of the day to stay on top of the many pitfalls that can cause one to fail in this business. That's one reason that I firmly believe in being extremely organized. That allows me to cover the numerous project-related items that need attention. Each construction job is like operating a separate business. It requires all of the effort and energy of running multiple businesses at the same time. That keeps the pace very fast and exciting. Of course, there is the personal reward at the end of the job of seeing the structure that you have built in its completed form.

"The business of construction is very interesting. When we took over the construction of the Los Angeles Mission in downtown Los Angeles from the contractor who had failed on the job, we realized the need for construction management and preconstruction services. We started to render those services in earnest during the past seven years, and it has proved to be the most interesting aspect of the work for me. We are able to assist the owners during the planning stages which helps save them money and time. That allows us to be very professional in an industry that is not very professional.

"The most rewarding part of the business is helping people meet their needs in planning and building structures for them. The business is very people oriented. You are constantly in dialogue with people—all day every day. You must be able to work with people. You need to have a team spirit. You perform your work along with many others. Construction is not a sole-person kind of job.

"The ability to communicate is a very necessary skill. Since my background and training are in communication and leading effectively by written and spoken word, I enjoy that part of the business very much also.

"The worst part of the business is working very hard to please a customer and receiving no appreciation for your effort. Since you are getting paid, they assume that you are making a lot of money and that means they can take advantage of you. People simply expect too much. Construction is not like making Swiss watches. Sometimes the more you try to please people, the more they expect.

"The other thing that I do not like is the dishonesty within the industry. Many companies do not care how they get a job or what it takes to convince an owner of their abilities. They will lie, cheat, and steal to persuade an owner to give them the work. Unfortunately, most owners do not find out about this until it's too late. So, often the owner just hears what he or she wants to hear. Many of these unscrupulous contractors could make as much money by being honest, but dishonesty seems to be endemic to the business.

"Salaries vary. Most project managers earn from $30,000 to $65,000 per year, depending on education and experience. Most superintendents make $45,000 to $60,000 per year, depending upon experience. My earnings are in the six figures, but keep in mind that I do the legal work for the company too.

Expert Advice

"My advice to someone who wants to get into the business is first to obtain a degree in construction management, engineering, or a similar field. Even a degree in communications or literature would be helpful. Any degree that helps train a person to be able to write and communicate will provide an advantage.

"You certainly need people skills, and you should be able to react to situations quickly. If you are interested in the field, you obviously need physical skills.

"If management is a goal, a graduate degree in law, business management, or finance would be good. There has been very little consolidation of businesses within the construction industry, mainly because the sophistication of management is not there. Once that changes, there will be a greater need for skilled management personnel with graduate degrees.

"At the entry level, you will need education and experience. The sooner you get that, the easier it will be to get started in the business. While going to school, you could work part time for a construction company or try to intern during the summer vacation periods."

INTERVIEW
Gene Formy-Duval
Licensed General Contractor

Gene Formy-Duval is self-employed and owns Sea Oats Realty & Construction in Southport, North Carolina. He is also a licensed real estate broker and has been working in the construction field since 1985.

How Gene Formy-Duval Got Started

"I grew up on a farm with a dad who believed 'If it's broken, you fix it yourself,' so I was exposed to construction at an early age. In 1963 I single-handedly remodeled my primary residence, a house dating back to the early 1920s. That more or less paved my career direction.

"I have one year of college in forestry management, with various specialty schools along the way. I moved my family to the small resort area of Southport, North Carolina, 13 years ago with the intention of concentrating on selling real estate. Then along came a buyer who needed a builder, and I said, 'Well, I can do that.'

"I passed the state exam, and the rest is history. While I prefer building spec homes, I occasionally go the custom gamut. As you might suspect, fewer people-interaction problems arise with spec houses."

What the Job's Really Like

"In home construction, 40-hour weeks are dreams that a builder imagines he'll have one day before he says farewell to this earthly life. Contractors put their reputation and license on the line every time they start construction; thus, their presence on a job site is felt by those subcontractors who tend to cut corners.

"Since I don't keep a crew of workers, I have to depend on contracting out phases of the work—plumbing, HVAC, insulation, drywall. Everyone claims to be a master carpenter, but the good ones are few. But even with all the headaches and problems encountered in building, I love it.

"A typical day early in a project consists of selecting the lot, requesting a percolation test by the county health department, and ordering the blueprints. Depending on the backlog of the health department, the time lapse between the percolation request and certification can be six weeks. This time allows the builder to get quotes from subcontractors. I do my own material takeoff, so I know going into a project what the final cost should be.

"With percolation certificate in hand, I get a permit to build, post the permit, and hire an excavator to clear the lot and build the pad where the house will stand.

"Next come the footings and piers on which the floor system will sit. Manufactured floor systems speed the process somewhat and cost about the same as stick-built plus labor. So, if time is a factor, go with the manufactured.

"From this point, you're working against time and weather. The dry-in process (outside wall studs, sheathing, and roofing) should be completed in about three days to avoid water damage to the already-in-place floor system. Windows and doors go in, and the inside finish work then proceeds at a much slower pace. Cabinets, carpet, hardwood flooring, and ceramic tile are the last finishes in place.

"After the drywall is finished, if this were a spec house, I would start advertising it for sale. This allows a buyer to contract with a builder, get a loan, and make those final selections in carpet color and so forth. The builder sets price-range limits on lighting, electrical points (receptacles), carpet, cabinets, and bathroom fixtures. Anything over and above limits is written as an amendment to the original contract and must be paid up front—the reasoning being that a builder doesn't want to get stuck with individual preferences that may not suit the next buyer in case the first buyer defaults. From my own experience, I wait until the bank approves the buyer's loan before I proceed with changes.

"Whether the house is spec or custom built, inspections from footings to finish are made by the building inspector and the lending institution.

"Profits can vary from 10 to 25 percent. Multibuilders get a material cost cut that occasional builders don't enjoy. The cost of the building site also factors into profit, as does the quality of subcontractor work.

"Good records must be kept for comparison purposes between houses and also because people tend to be 'sue happy' these days.

"Would I become a builder if I could replan my life? You bet!"

Expert Advice

"Knowledge about the business of building houses can be gained in libraries, bookstores, and hands-on. Many crafts people don't go that extra mile to become professionals because studying the code book, which covers all facets of building, and passing the state exam require the basic elements—reading, arithmetic, and comprehension. You shouldn't go into home building with dollar signs in your eyes. The look of satisfaction on the home owner's face means more work for the builder down the road. In this business, word of mouth can mean the difference between success and failure.

"You must be willing to keep long hours, deal with distributors over delays, and keep up with the material on site so it doesn't walk off. Even the most even-tempered people lose their cool occasionally, but you must practice biting your tongue when it's in danger of getting chopped off. The field is wide open for those who are serious about doing this kind of work."

INTERVIEW
Joy Formy-Duval
Project Scheduler

Joy Formy-Duval works for Miller, a family-owned construction/ development business with offices in Wilmington, Raleigh, and Charlotte, North Carolina. She has been in the construction business for 12 years and started as Miller's scheduler in 1994.

How Joy Formy-Duval Got Started

"My background is accounting, but when I moved to the small resort community of Southport, North Carolina, the local nuclear plant was the only employer to pay decent wages, so I became a contract technical assistant working with engineers on long-term projects. This eventually took me to Carolina Power & Light's corporate office in Raleigh for four years. While I never grew to like Porta-Johns, I did enjoy my work. After eight years of doing this, I'd had enough living out of a suitcase, and I came home . . . again . . . faced with the possibility of low wages or not working.

"I applied to Miller, an hour's drive away, although they weren't advertising at the time. What did I have to lose? I got hired on the spot. Little did I know that their present scheduler had just quit from burnout.

"I have two years of college, 30-plus hours of continuing ed, and an associate degree in business administration and one in real estate.

"My training came on the job and from being married to a residential builder for many years.

"The creativity of seeing major jobs from design through completion attracted me to this field."

What the Job's Really Like

"Scheduling is the life of a project tracked on paper—in the beginning, an overview of what we propose to a prospective client. Companies, large and small, have project schedulers to plot and keep up with the jobs. Clients demand it before going into a contract with a builder. The client wants to know the exact duration from start to finish because the bottom line is 'How soon can I move in and start business?'

"After the contract between the builder and the client is signed, a more detailed schedule emerges—one that reflects activity duration from design to permitting to owner occupancy.

"My job is to interact with the project manager in charge of the project and the on-site superintendent in tracking each activity. I have to show them weather, material, or subcontractor delays, as well as how on schedule we are. One of Miller's biggest selling tools in getting new contracts is the ability to get the job done in the time we said we'd do it. The bottom line for the customer is 'How soon can I use it?'

"Depending on the dollar size of the job (the company I'm with has projects in the $50 million range), a schedule may have more than a hundred activities, ranging from permitting to finished product.

"I'm in a position more often than not held by a male, but today more and more women are getting into the construction field and doing a great job. My company had never had a female scheduler until I joined them four years ago. I came in with 10 years' experience with the nuclear field. But I've found there is still wage discrimination between the sexes. After four years, my $26,000 annual salary is still lower than the former male scheduler, who had no more time invested than I do now.

"Construction, whether it is residential, institutional, commercial, or industrial, requires that a certain line of progress takes place called 'critical path.' That means the footings can't be poured before the pad is built, exterior skin can't be put on before the studs are in place, and painting can't take place before the drywall finish is complete . . . just to name a few.

"I have learned from experience, though, that a roof can be put on before the slab is poured. A couple of years back on a manufacturing facility, daily rain delays pushed the schedule back. After two weeks without a letup, we put the roof on and then piped in concrete for the slab (flooring). Most contracts have heavy penalty clauses for going beyond the client occupancy date set originally by the client and builder, so the amount of rainfall and temperature are checked three times daily on each job. Concrete forms too many ice crystals if poured when the temperature is below 32 degrees Fahrenheit.

"Most weeks are 40 hours. A few are boring; many are hairy because so many people are involved in our large operation, and I have an average of 25 ongoing projects that I must track.

"The job itself entails entering lots and lots of data into a scheduling software program and turning that data into a plot printed on an HP 650C plotter. One has to have a knowledge of construction to know the line of progression required for a structure to 'rise from the barren land.'

"The downside of the job is its repetitious nature. The rewarding side is seeing my name on the schedules hanging on some very prestigious walls. My credentials warrant a page in the prospectus booklet given to clients, and surprisingly, the clients want to meet the person tracking their project."

Expert Advice

"If one doesn't have a natural aptitude for the building process and the software, such things can be learned at community colleges and universities. Interacting with the field people keeps the job from becoming boring and routine. Staring at a computer screen eight hours a day, plotting slabs on grade, underground, rough-in plumbing, and the like requires the maturity to set boredom aside and get the task finished, because someone out there is waiting for it.

"Would I recommend becoming a scheduler? Yes, I would, but then I like having a hand in the birth of a new structure, working around obstacles, knowing that 10 months down the road we'll add another satisfied customer to our list."

FOR MORE INFORMATION

For information about a career as a construction manager, contact:

American Institute of Constructors
466 94th Avenue North
St. Petersburg, FL 33702

Associated Builders and Contractors
1300 North 17th Street
Rosslyn, VA 22209

Associated General Contractors of America, Inc.
1957 E Street NW
Washington, D.C. 20006-5199

Construction Management Association of America
7918 Jones Branch Drive, Suite 540
McLean, VA 22102

Information on the accreditation requirements for construction science and management programs is available from:

American Council for Construction Education
1300 Hudson Lane, Suite 3
Monroe, LA 71201-6054

CHAPTER 4 Cost Estimators

EDUCATION
B.A./B.S. Preferred

$$$ SALARY
$17,000 to $75,000

OVERVIEW

Accurately predicting the cost of future projects is vital to the survival of any business. Cost estimators develop cost information for owners or managers to use in making bids for contracts, in determining if a new product will be profitable, or in determining which of a firm's products are making a profit.

Regardless of the industry they work in, estimators compile and analyze data on all the factors that can influence costs, such as materials, labor, location, and special machinery requirements, including computer hardware and software. Job duties vary widely depending on the type and size of the project. Estimators working in the construction industry and manufacturing businesses have different methods of and motivations for estimating costs.

On a large construction project, for example, the estimating process begins with the decision to submit a bid. After reviewing the architect's drawings and specifications, the estimator visits the site of the proposed project. The estimator needs to gather information on access to the site and availability of electricity, water, and other services, as well as surface topography and drainage. If the project is a remodeling or renovation job, the estimator might consider the need to control noise and dust and schedule work in order to accommodate occupants of the building. The information developed during the site visit generally is recorded in a signed report that is made part of the final project estimate.

After the site visit is completed, the estimator determines the quantity of materials and labor that the firm will have to furnish. This process, called the quantity survey, or "takeoff," is completed by filling out standard estimating forms that provide spaces for the entry of dimensions, number of units, and other information. A cost estimator working for a general contractor, for example, will estimate the costs of all items that the contractor must provide. Although subcontractors will estimate their costs as part of their own bidding process, the general contractor's cost estimator often analyzes bids made by subcontractors as well. Also during the takeoff process, the estimator must make decisions concerning equipment needs, sequence of operations, and crew size. Allowances for the waste of materials, inclement weather, shipping delays, and other factors that may increase costs are likewise incorporated in the takeoff.

On completion of the quantity surveys, a total project cost summary is prepared by the chief estimator that includes the cost of labor, equipment, materials, subcontracts, overhead, taxes, insurance, and markup, and any other costs that may affect the project. The chief estimator then prepares the bid proposal for submission to the developer.

Construction cost estimators also may be employed by the project's architect or owner to estimate costs or track actual costs relative to bid specifications as the project develops. In large construction companies that employ more than one estimator, it is common practice for estimators to specialize. For instance, one person may estimate only electrical work, while another may concentrate on excavation, concrete, and forms.

In manufacturing and other firms, cost estimators generally are assigned to the engineering or cost department. The estimators' goal in manufacturing is to accurately allocate the costs associated with making products. The job may begin when management requests an estimate of the costs associated with a major redesign of an existing product or the development of a new product or production process. When estimating the cost of developing a new product, for example, the estimator works with engineers, first reviewing blueprints or conceptual drawings to determine the machining operations, tools, gauges, and materials that would be required for the job.

The estimator then prepares a parts list and determines whether it is more efficient to produce or to purchase the parts. To do this, the estimator must initiate inquiries for price information from potential suppliers. The next step is to determine the cost of manufacturing each component of the product. Some high-technology products require a tremendous amount of computer programming during the design phase. The cost of software development is one of the fastest-growing and most difficult activities to estimate; some cost estimators now specialize in estimating only computer software development and related costs.

The cost estimator then prepares time-phase charts and learning curves. Time-phase charts indicate the time required for tool design and fabrication, tool "debugging"—finding and correcting all problems—manufacturing of parts, assembly, and testing. Learning curves graphically represent the rate at which performance improves with practice. These curves are commonly called "problem-elimination" curves because many problems such as engineering changes, rework, parts shortages, and lack of operator skills diminish as the number of parts produced increases, resulting in lower unit costs.

Using all of this information, the estimator then calculates the standard labor hours necessary to produce a predetermined number of units. Standard labor hours are then converted to dollar values, to which are added factors for waste, overhead, and profit to yield the unit cost in dollars. The estimator then compares the cost of purchasing parts with the firm's cost of manufacturing them to determine which is cheaper.

Computers are widely used because cost estimating may involve complex calculations and require advanced mathematical techniques. For example, to undertake a parametric analysis, a process used to estimate project costs on a per-unit basis subject to the specific requirements of a project, cost estimators use a computer database containing information on costs and conditions of many other similar projects. Although computers cannot be used for the entire estimating process, they can relieve estimators of much of the drudgery associated with routine, repetitive, and time-consuming calculations. Computers also are used to produce all of the necessary documentation with the help of basic word-processing and spreadsheet software. This leaves estimators with more time to study and analyze projects and can lead to more accurate estimates. (More detailed information on various cost estimating techniques is available from the organizations listed at the end of the chapter.)

Although estimators spend most of their time in an office, construction estimators must make frequent visits to work sites that are dirty and cluttered with debris. Likewise, estimators in manufacturing must spend time on the factory floor, where it can be hot, noisy, and dirty.

Cost estimators usually operate under pressure, especially when facing deadlines. Inaccurate estimating can cause a firm to lose out on a bid or lose money on a job that proves to be unprofitable. Although estimators normally work a 40-hour week, overtime is often required. In some industries, frequent travel between a firm's headquarters and its subsidiaries or subcontractors also may be required.

TRAINING

Entry requirements for cost estimators vary significantly by industry. In the construction industry, employers prefer applicants with a thorough knowledge of construction materials, costs, and procedures in areas ranging from heavy construction to electrical work, plumbing systems, and masonry work. Most construction estimators have considerable experience as a construction craft worker or manager. Individuals who combine this experience with some postsecondary training in construction estimating, or with a bachelor's or associate degree in civil engineering, architectural drafting, or building construction, have a competitive edge in landing jobs.

In manufacturing industries, employers prefer to hire individuals with a degree in engineering, science, operations research, mathematics, or statistics, or in accounting, finance, business, or a related subject. In high-technology industries, great emphasis is placed on experience involving quantitative techniques.

Cost estimators should have an aptitude for mathematics, be able to quickly analyze, compare, and interpret detailed and sometimes poorly defined information, and be able to make sound and accurate judgments based on this knowledge. Assertiveness and self-confidence in presenting and supporting conclusions are important as well. Cost estimators should also be familiar with computers and their application to the estimating process, including word-processing and spreadsheet packages used to produce necessary documentation. In some instances, familiarity with special estimation software or programming skills may be useful.

Regardless of their background, estimators receive much training on the job. Working with an experienced estimator, they become familiar with each step in the process. Those with no experience reading construction specifications or blueprints first learn that aspect of the work. They then may accompany an experienced estimator to the construction site or shop floor, where they observe the work being done, take measurements, or perform other routine tasks. As they become more knowledgeable, estimators learn how to tabulate quantities and dimensions from drawings and how to select the appropriate material prices.

Many colleges and universities include cost estimating as part of the curriculum in civil engineering, industrial engineering, and construction management or construction engineering technology. Courses and programs in cost estimating techniques and procedures are also offered by many technical schools, junior colleges, and universities.

In addition, cost estimating is a significant part of master's degree programs in construction management offered by many colleges and universities. Organizations that represent cost estimators, such as the American Association of Cost Engineers International (AACE) and the Society of Cost Estimating and Analysis, also sponsor educational programs. These programs help students, estimators-in-training, and experienced estimators stay abreast of changes affecting the profession.

Voluntary certification can be valuable to cost estimators because it provides professional recognition of the estimator's competence and experience. Both AACE International and the Society of Cost Estimating and Analysis administer certification programs. To become certified, estimators generally must have between three and seven years of estimating experience and must pass both a written and an oral examination. In addition, certification requirements may include publication of at least one article or paper in the field.

For most estimators, advancement takes the form of higher pay and prestige. Some move into management positions, such as project manager for a construction firm or manager of the industrial engineering department for a manufacturer. Others may go into business for themselves as consultants, providing estimating services for a fee to government or construction and manufacturing firms.

JOB OUTLOOK

Cost estimators hold about 179,000 jobs, primarily in construction industries. Others can be found in manufacturing industries. Some cost estimators also work for engineering and architectural services firms, business services firms, and throughout a wide range of other industries. Construction, operations research, production control, cost, and price analysts who work for government agencies also may do significant amounts of cost estimating in the course of their regular duties.

Cost estimators work throughout the country, usually in or near major industrial, commercial, and government centers, and in cities and suburban areas undergoing rapid change or development.

Overall employment of cost estimators is expected to grow about as fast as the average for all occupations through the year 2005 as the levels of construction and manufacturing activity increase in line with economic growth. However, even when construction and manufacturing activity decline, there should always remain a demand for cost estimators to accurately predict costs in all areas of business. Some job openings will also arise from the need to replace workers who transfer to other occupations or who leave the labor force altogether.

Growth of the construction industry, where more than 60 percent of all cost estimators are employed, will be the driving force behind the rising demand for these workers. The fastest-growing sectors of the construction industry are expected to be special trade contractors and those associated with heavy construction and spending on the nation's infrastructure. Construction and repair of highways, streets, and bridges, and construction of more subway systems, airports, water and sewage systems, and electric power plants and transmission lines will stimulate demand for many more cost estimators. Job prospects in construction should be best for workers with a degree in construction management, engineering, or architectural drafting, or who have substantial experience in various phases of construction or a specialty craft area.

Employment of cost estimators in manufacturing should remain relatively stable as firms continue to use their services to identify and control operating costs. Experienced estimators with degrees in engineering, science, mathematics, business administration, or economics and who have computer expertise should have the best job prospects in manufacturing.

SALARIES

Salaries of cost estimators vary widely by experience, education, size of firm, and industry. According to the data available, most starting salaries in the construction industry for cost estimators with limited training were between about $17,000 and $21,000 a year in 1994.

College graduates with degrees in fields such as engineering or construction management that provide a strong background in cost estimating could start at about $30,000 annually or more. Highly experienced cost estimators earn $75,000 a year or more. Starting salaries and annual earnings in the manufacturing sector usually are somewhat higher.

RELATED FIELDS

Workers who quantitatively analyze information in a similar capacity include appraisers, cost accountants, cost engineers, economists, evaluators, financial analysts, loan officers, operations research analysts, underwriters, and value engineers.

INTERVIEW
Lee Sullivan Hill
Cost Estimator

Lee Sullivan Hill works for Turner Construction Company, an international construction management and general contracting firm headquartered in New York City. Lee worked in the Washington, D.C., office for seven years, then in Connecticut for two years, and is now in Chicago. She received an A.B. degree in engineering from Lafayette College in Easton, Pennsylvania.

How Lee Sullivan Hill Got Started

"As a child, I had never in my life met an engineer. I did not even know what one was. I did, however, love architecture. My grandmother's father had been an 'architect' (so said my grandmother—I learned later that he had actually been a mechanical engineer!), and time spent on walks with her taught me to look for mansard roofs and Doric columns by age five. But life went on, and I went off to college hoping to get A's and go on to become a veterinarian.

"Fast-forward to my freshman year. My roommate was an engineering student. I loved her homework assignments—they looked so interesting. I had always been good at math and interested in architecture, so I decided (three weeks into my sophomore year!) to switch into engineering. The A.B. engineering program at Lafayette allowed me to take civil engineering courses such as structures and design of small dams as well as art classes, history, English, and French. I loved it!

"I had planned to go to graduate school for architecture, until fate stepped in again. My boyfriend (husband-to-be) graduated two years before I did and went to work with Turner Construction. I loved hearing about his job—building a dormitory at Gallaudet College in D.C. By the time May rolled around, I had found my niche. I got offers from other construction firms, but I chose Turner.

"I started with Turner right out of college in June of 1980. My first assignments were on construction job sites. I was assigned to the Cost Estimating Department (in Washington, D.C.) in 1982. At the end of 1986, I took maternity leave. In 1993 I returned to my estimating duties on a part-time, hourly basis. I have done just one estimate for Turner since moving to Chicago in April of 1996 because my 'other' career, writing children's books—engineering books, I might add—has accelerated.

"The training at Turner took me through work on job sites, but I naturally gravitated toward estimating. Even on my first job where I laid out concrete for a water treatment plant, I did takeoff and labor studies. My work was neat, organized, and thorough. I think the company pegged me for estimating by the time I was there six months.

"The training within estimating was mostly one-on-one with my boss. The department had four people in it and included the chief, so there was plenty of time for this method.

"My natural talents for organizing information made the learning simple. In 1982, when I started, the computers were not yet a factor. (Actually, it became my job in 1985–1986 to computerize the department and provide training to the other estimators.) My boss started me out doing counting of door frames and windows. Pretty soon, I was taking on small projects all on my own.

"Turner Construction also has a formalized training program of seminars that contributed greatly to my knowledge. Reading trade journals and talking with subcontractors about their work added finishing touches."

What the Job's Really Like

"First of all, all estimators spend time looking at drawings and specifications for upcoming construction projects. They study the plans, read the specs, take notes on unusual (maybe expensive) details, and make lists of questions and qualifications. They think about and envision the finished building: What problems will arise during construction? What has the architect left out of the design that will be needed to finish the work? Can we change some details that will save the owner money and still provide a quality building? Are there some details that we know cause major problems (such as rainwater seeping through bad flashing details) that we should red-flag?

"Second, all estimators count. They use digitizers that count with computer programs, or they count with their fingers, but all estimators must figure out how many square feet, cubic yards, linear feet . . . how many of every item and how much and how long. This takeoff must be done in an organized fashion. And it must be checked and rechecked (oops—you didn't leave out a floor, did you?).

"Third, all estimators assign prices to the work. For every item they counted, they assign a price. This is more complicated than you may imagine. It's not just running to a fat book and pulling out a price. Every job has specifics that may make prices higher or lower. Will the work go on in winter (need extra money for heat)? Are new labor contracts coming up that will change the costs? How about insurance requirements? All the details must be compared. Subcontractors called on the phone can offer advice. Discussions with the owner and architect clarify unknowns.

"This brings us to the most important task of all: communication. All the organized notes and careful pricing and checking of work and rechecking of work are nothing without communication. Final reports, lump-sum bids, and meetings with owners before construction starts are all critical components of an estimator's job.

"On large projects, the estimate is broken into pieces and assigned to several estimators. The lead estimator must communicate with everyone and pull the information together without dropping a piece that might cost big money later. The ability to communicate is hard to train into people. Companies look for this quality in all of their hires.

"To sum up, estimating is a fun, stimulating, exciting process that involves seemingly opposite activities—independent activities such as thinking and counting, and communication activities such as contacting subcontractors, owners, architects, and people within your own company. One thing it never is: boring. Absolutely never. Well, counting door frames isn't exciting, but you get to imagine what the frames will look like when the building is complete!

"Part-time estimating is super because I work on a job, get a check, and go back to my life as mom and writer.

"As far as when I worked full time, I liked the interaction with people, the planning/cost-cutting of preconstruction when we worked as a team with the architect, builder, and owner.

"I loved seeing plans for new buildings, imagining their beauty, imagining the smell of the concrete foundations freshly poured, the wood floors freshly varnished.

"I can't say there was anything I hated. Except losing. Sometimes you'd bust your gut for a month on a project, put in a price/bid, and find that another contractor got the job for a million less. Killer.

"As far as salaries go, now, as a part-time worker, I am paid hourly. This is not a normal situation in this field. Turner happens to be flexible and has found it to be a way to keep good women engineers who want to balance family and career. I receive $25 per hour. They also give an end-of-the-year bonus proportional to your salary. A full-time estimator would be paid on a salary and make anywhere from $40,000 to $100,000—it really depends on years of experience and level."

Expert Advice

"Study civil engineering, construction management, or architecture in college. Get as much work experience related to the field as you can during the summers. Even being a laborer on a job site counts as construction experience when you look for your job at graduation. And hone your communication (both written and oral) skills. They will serve you better than anything you could write in your resume."

FOR MORE INFORMATION

Information about career opportunities, certification, and educational programs in cost estimating in the construction industry may be obtained from:

AACE International
209 Prairie Avenue, Suite 100
Morgantown, WV 26505

Professional Construction Estimators Association of America
P.O. Box 11626
Charlotte, NC 28220-1626

Similar information about cost estimating in government, manufacturing, and other industries is available from:

Society of Cost Estimating and Analysis
101 South Whiting Street, Suite 201
Alexandria, VA 22304

5

Bricklayers and Stonemasons

EDUCATION
H.S. Preferred

SALARY
$13,000 to $48,000

OVERVIEW

Bricklayers and stonemasons work in closely related trades that produce attractive, durable surfaces and structures. The work they perform varies in complexity, from laying a simple masonry walkway to installing the ornate exterior of a high-rise building.

Bricklayers build walls, floors, partitions, fireplaces, chimneys, and other structures with brick, precast masonry panels, concrete block, and other masonry materials. Some specialize in installing firebrick linings in industrial furnaces.

Stonemasons build stone walls as well as set stone exteriors and floors. They work with two types of stone: natural cut, such as marble, granite, and limestone, and artificial stone made from concrete, marble chips, or other masonry materials. Stonemasons usually work on structures such as houses of worship, hotels, and office buildings.

In putting up a wall, bricklayers build the corners of the structure first. Because of the necessary precision, these corner leads are very time-consuming to erect and require the skills of the most experienced bricklayers on the job. After the corner leads are complete, less experienced bricklayers fill in the wall between the corners, using a line from corner to corner to guide each course, or layer, of brick. Because of the expense associated with building corner leads, an increasing number of bricklayers are using corner poles, also called masonry guides, that enable them to build the entire wall at the same time. They fasten the corner posts or poles in a plumb position to define the wall line and stretch a line between them. The line serves as a guide for each course of brick. Bricklayers then spread a bed of mortar (a cement, sand, and water mixture) with a trowel (a flat, bladed metal tool with a handle), place the brick on the mortar bed, and then press and tap it into place. As blueprints specify, they either cut bricks with a hammer and chisel or saw them to fit around windows, doors, and other openings. Mortar joints are finished with jointing tools for a sealed, neat, and uniform appearance. Although bricklayers generally use steel supports or "lintels" at window and door openings, they sometimes build brick arches that support and enhance the beauty of the brickwork.

Bricklayers are assisted by hod carriers, or helpers, who carry brick and other materials, mix mortar, and set up and move the scaffolding.

Stonemasons often work from a set of drawings in which each stone has been numbered for identification. Helpers may locate and carry the prenumbered stones to the masons. A derrick operator using a hoist may be needed to lift large pieces into place.

When building a stone wall, masons set the first course of stones into a shallow bed of mortar. They align the stones with wedges, plumb lines, and levels, and adjust them into position with a hard rubber mallet. Masons build the wall by alternating layers of mortar and courses of stone. As the work progresses, they remove the wedges and fill the joints between stones and use a pointed metal tool, called a tuck-pointer, to smooth the mortar to an attractive finish. To hold large stones in place, stonemasons attach brackets to the stone and weld or bolt them to anchors in the wall. Finally, masons wash the stone with a cleansing solution to remove stains and dry mortar.

When setting stone floors, which often consist of large and heavy pieces of stone, masons first trowel a layer of damp mortar over the surface to be covered. Using crowbars and hard rubber mallets for aligning and leveling, they then set the stone in the mortar bed. To finish, workers fill the joints and wash the stone slabs.

Masons use a special hammer and chisel to cut stone. They cut it along the grain to make various shapes and sizes. Valuable pieces often are cut with a saw that has a diamond blade. Some masons specialize in setting marble, which, in many respects, is similar to setting large pieces of stone. Bricklayers and stonemasons also repair imperfections and cracks or replace broken or missing masonry units in walls and floors.

Most nonresidential buildings are now built with prefabricated panels made of concrete block, brick veneer, stone, granite, marble, tile, or glass. In the past, bricklayers did mostly interior work, such as block partition walls and elevator shafts. Now they must be more versatile and work with many materials. For example, bricklayers now install lighter-weight insulated panels used in new skyscraper construction.

Refractory masons are bricklayers who specialize in installing firebrick and refractory tile in high-temperature boilers, furnaces, cupolas, ladles, and soaking pits in industrial establishments. Most refractory masons work in steel mills, where molten materials flow on refractory beds from furnaces to rolling machines.

Bricklayers and stonemasons usually work outdoors. They stand, kneel, and bend for long periods and often have to lift heavy materials. Common hazards include injuries from tools and falls from scaffolds, but these can be avoided when proper safety practices are followed.

TRAINING

Most bricklayers and stonemasons pick up their skills informally by observing and learning from experienced workers. Many get training in vocational education schools. The best way to learn these skills, however, is through an apprenticeship program, which generally provides the most thorough training.

Individuals who learn the trade on the job usually start as helpers, laborers, or mason tenders. They carry materials, move scaffolds, and mix mortar. When the opportunity arises, they are taught to spread mortar, lay brick and block, or set stone. As they gain experience, they make the transition to full-fledged craft workers. The learning period generally lasts much longer than an apprenticeship program, however.

Apprenticeships for bricklayers and stonemasons usually are sponsored by local contractors or by local union-management committees. The apprenticeship program requires three years of on-the-job training in addition to a minimum 144 hours of classroom instruction each year in subjects such as blueprint reading, mathematics, layout work, and sketching.

Apprentices often start by working with laborers, carrying materials, mixing mortar, and building scaffolds. This period generally lasts about a month and familiarizes them with job routines and materials. Next, they learn to lay, align, and join brick and block. Apprentices also learn to work with stone and concrete. This enables them to be certified to work with more than one masonry material.

Applicants for apprenticeships must be at least 17 years old and in good physical condition. A high school education is preferable, and courses in mathematics, mechanical drawing, and shop are helpful. The International Masonry Institute, a division of the International Union of Bricklayers and Allied Craftsmen, operates training centers in several large cities that help job seekers develop the skills they will need to successfully complete the formal apprenticeship program.

Experienced workers can advance to supervisory positions or become estimators. They also can open contracting businesses of their own.

JOB OUTLOOK

Bricklayers and stonemasons hold about 142,000 jobs nationwide. The vast majority are bricklayers. Workers in these crafts are employed primarily by special trade, building, or general contractors. They work throughout the country but, like the general population, are concentrated in metropolitan areas.

Nearly three of every ten bricklayers and stonemasons are self-employed. Many of the self-employed specialize in contracting on small jobs such as patios, walks, and fireplaces.

Job opportunities for skilled bricklayers and stonemasons are expected to be good as the growth in demand outpaces the supply of workers trained in this craft.

Employment of bricklayers and stonemasons is expected to grow about as fast as the average for all occupations through the year 2005, and additional openings will result from the need to replace bricklayers and stonemasons who retire, transfer to other occupations, or leave the trades for other reasons. The pool of young workers available to enter training programs will also be increasing slowly, and many in that group are reluctant to seek training for jobs that may be strenuous and have uncomfortable working conditions.

Population and business growth will create a need for new factories, schools, hospitals, offices, and other structures, increasing the demand for bricklayers and stonemasons. Also stimulating demand will be the need to restore a growing stock of old masonry buildings, as well as the increasing use of brick for decorative work on building fronts and in lobbies and foyers.

Brick exteriors should continue to be very popular as the trend continues toward more durable exterior materials requiring less maintenance. Employment of bricklayers who specialize in refractory repair will decline, along with employment in other occupations in the primary metal industries.

Employment of bricklayers and stonemasons, like that of many other construction workers, is sensitive to changes in the economy. When the level of construction activity falls, workers in these trades can experience periods of unemployment.

SALARIES

Median weekly earnings for bricklayers and stonemasons were about $484 in 1996. The middle 50 percent earned between $345 and $624 weekly. The highest 10 percent earned more than $926 weekly, and the lowest 10 percent earned less than $247. Earnings for workers in these trades may be reduced on occasion because poor weather and downturns in construction activity limit the time they can work.

In each trade, apprentices or helpers usually start at about 50 percent of the wage rate paid to experienced workers. This increases as they gain experience.

Some bricklayers and stonemasons are members of the International Union of Bricklayers and Allied Craftsmen.

RELATED FIELDS

Bricklayers and stonemasons combine a thorough knowledge of brick, concrete block, stone, and marble with manual skill to erect very attractive yet highly durable structures. Workers in occupations with similar skills include concrete masons, plasterers, terrazzo workers, and tile setters.

INTERVIEW
David McCafferty
Brick Mason

David McCafferty is a partner in M.R. Construction, Inc., in Warren, Ohio. He has been working in this field since 1994.

How David McCafferty Got Started

"It was by coincidence that I got started. Mac, my dad's neighbor, was a masonry contractor, and one day I went over to ask if he was hiring. He said, no, he had two laborers already, but I insisted on leaving a resume. Next day, he called me. His two laborers hadn't shown up for work for the second time, and he fired them.

"I had to work hard, but working with Mac made me want to do it. I gave him 110 percent every day for almost nine years. Then he told Russ, my partner in M.R. Construction, and I that he was going to retire, and he gave us his business. He let us borrow all his equipment, and he worked for us part time to make sure we got off on the right foot. And he backed us up with the general contractors, telling them he would guarantee our work.

"I can't say enough about Mac. He did a lot for me and Russ. I'm not sure we could have started this business without his help. Equipment costs a lot to get started. After two years, he sold the equipment to us for almost nothing because he wanted us to succeed at it, he said."

What the Job's Really Like

"I have a partner, and we own the company. We get paid by the block or brick, so, the faster we get it down, the more money we make. But it's hard work, and there's a lot of responsibilities.

"We have to make sure that everyone is working safely, yet being productive. It's backbreaking work. If it's not 94 degrees, it's 30 degrees or raining or snowing or thundering. It's always something—but I love it.

"You have to make sure that the foundation you are going to build is squared, so, before material is delivered, we check the pins—the nails in the concrete footer where the corners of the house go. If it's wrong, the mason is blamed and no one else.

"Then I have to figure out how many blocks it's going to take to do the job and correct any errors on the blueprint. I call in or fax the order to the block company to have the material delivered. Scheduling jobs is the hardest part. You can't just say that it's going to take three or four or six days to do a job. You might get rain for three days, then the banks cave in (that's the dirt going up alongside the basement), or the home owner might make a change, and that can slow you down. Then you get a domino effect: each delay affects the next task.

"We have to hire employees, and we try to pay them decently. I go through about six employees a year. Then we have to train someone new all over again. And this slows you down.

"I try to get about 300 blocks laid a day, but that's hard to do if the job is all cut up (has a lot of corners) or if I'm trying to train a new employee.

"We expect everyone, including ourselves, to do nice, clean work. We clean up after ourselves and scrape the walls off and the footers, making it easier on the next guy.

"The work can be fulfilling, if you find satisfaction in finishing a job. Russ and I like to listen to the radio. We sing; we joke around and just have fun.

"We work about 50 hours a week in the summer, but you may not work at all in the winter, so we pay unemployment on ourselves. Your hours depend on how much work is out there. You also have to deal with how much competition you have and how hard you want to work. How well you do is dependent on how good you are and how much you charge.

"I like looking at the job after it's down and saying, 'We did that.' Taking the check to the bank is nice, too. I like working outside, also. Someone telling me that I do good work is very rewarding. I especially enjoy doing fancy, detail brickwork, where you have arches and stonework and the clients are willing to pay for it.

"I dislike having to deal with home owners. I'd rather be given a set of blueprints and turned loose. I dislike having to tear down something I built because there was a mistake made. Other downsides are having to chase people for my money or bickering with contractors over the price per block because they want to cheat me out of a few dollars.

"I also don't like it when I can't relax and have fun on the job because I have to watch employees every minute to make sure they're not doing something wrong. I don't like having to pay them to do it again the right way.

"We structure our salary on what we could afford to pay ourselves. Don't expect a whole lot starting out. We worked two months without pay till we could afford to pay ourselves. We were finally able to pay ourselves retroactively after we got going and built up some capital. But our employees always come first. They never had to wait for their checks. That's very important."

Expert Advice

"You have to be dedicated to what you want to do. Then find the market for that—the contractors who do the kind of work you are looking to do. And find a good accountant.

"You have to be able to think on your feet. You need to be able to look at something and figure out how much to charge and how much it's going to cost to have it built. All it takes is underestimating one big job, and you're going to be set back and set back hard.

"Stay flexible. Never commit yourself to one contractor or keep all your eggs in one basket. What if that contractor goes under and can't pay you? Have more than one contractor to work for.

"Find a good mason to start with, and learn from that person how to do things—and also learn from that person's mistakes."

FOR MORE INFORMATION

For details about apprenticeships or other work opportunities in these trades, contact local bricklaying, stonemasonry, or marble setting contractors; a local of the union; a local joint union-management apprenticeship committee; or the nearest office of the state employment service or state apprenticeship agency.

For general information about the work of either bricklayers or stonemasons, contact:

International Union of Bricklayers and Allied Craftsmen
International Masonry Institute Apprenticeship and Training
815 15th Street NW, Suite 1001
Washington, D.C. 20005

Information about the work of bricklayers also may be obtained from:

Associated General Contractors of America, Inc.
1957 E Street NW
Washington, D.C. 20006-5199

Brick Institute of America
11490 Commerce Park Drive
Reston, VA 22091

Home Builders Institute
National Association of Home Builders
1201 15th Street NW
Washington, D.C. 20005

National Concrete Masonry Association
2302 Horse Pen Road
Herndon, VA 22071

CHAPTER 6 Carpenters

EDUCATION
H.S. Preferred

$$$ SALARY
$14,000 to $45,000

OVERVIEW

Carpenters are involved in many different kinds of construction activities. They cut, fit, and assemble wood and other materials in the construction of buildings, highways and bridges, docks, industrial plants, boats, and many other structures.

Their duties vary depending on their type of employer. A carpenter employed by a special trade contractor, for example, may specialize in one or two activities, such as setting forms for concrete construction or erecting scaffolding. However, a carpenter employed by a general building contractor may perform many tasks, such as framing walls and partitions, putting in doors and windows, hanging kitchen cabinets, and installing paneling and tile ceilings.

Local building codes often dictate where certain materials can be used, and carpenters have to know these requirements. Each carpentry task is somewhat different, but most tasks involve the same basic steps. Working from blueprints or instructions from supervisors, carpenters first do the layout—measuring, marking, and arranging materials. They then cut and shape wood, plastic, ceiling tile, fiberglass, or drywall, using hand and power tools, such as chisels, planes, saws, drills, and sanders, and then join the materials with nails, screws, staples, or adhesives. In the final step, they check the accuracy of their work with levels, rules, plumb bobs, and framing squares and make any necessary adjustments.

When prefabricated components, such as stairs or wall panels, are used, the carpenter's task is somewhat simpler because it does not require as much layout work or the cutting and assembly of as many pieces. These components are designed for easy and fast installation and can generally be installed in a single operation.

Carpenters employed outside the construction industry do a variety of installation and maintenance work. They may replace panes of glass, ceiling tiles, and doors, as well as repair desks, cabinets, and other furniture. Depending on the employer, they may install partitions, doors, and windows; change locks; and repair broken furniture. In manufacturing firms, carpenters may assist in moving or installing machinery.

As in other building trades, carpentry work is sometimes strenuous. Prolonged standing, climbing, bending, and kneeling often are necessary. Carpenters risk injury from slips or falls, from working with sharp or rough materials, and from the use of sharp tools and power equipment. Many carpenters work outdoors.

Some carpenters change employers each time they finish a construction job. Others alternate between working for a contractor and working as contractors themselves on small jobs.

TRAINING

Carpenters learn their trade through on-the-job training and through formal training programs. Some pick up skills informally by working under the supervision of experienced workers. Many acquire skills through vocational education. Others participate in employer training programs or apprenticeships.

Most employers recommend an apprenticeship as the best way to learn carpentry. Because the number of apprenticeship programs is limited, however, only a small proportion of carpenters learn their trade through these programs. Apprenticeship programs are administered by local joint union-management committees of the United Brotherhood of Carpenters and Joiners of America and the Associated General Contractors of America, Inc., or the National Association of Home Builders.

Training programs are administered by local chapters of the Associated Builders and Contractors and by local chapters of the Associated General Contractors of America, Inc. These programs combine on-the-job training with related classroom instruction. Apprenticeship applicants generally must be at least 17 years old and meet local requirements. For example, some union locals test an applicant's aptitude for carpentry. The length of the program, usually about three to four years, varies with the apprentice's skill.

On the job, apprentices learn elementary structural design and become familiar with common carpentry jobs such as layout, form building, rough framing, and outside and inside finishing. They also learn to use the tools, machines, equipment, and materials of the trade. Apprentices receive classroom instruction in safety, first aid, blueprint reading and freehand sketching, basic mathematics, and different carpentry techniques. Both in the classroom and on the job, they learn the relationship between carpentry and the other building trades.

Informal on-the-job training usually is less thorough than an apprenticeship. The degree of training and supervision often depends on the size of the employing firm. A small contractor who specializes in home building may provide training only in rough framing. In contrast, a large general contractor may provide training in several carpentry skills. Although specialization is becoming increasingly common, it is important to try to acquire skills in all aspects of carpentry to have the flexibility to be able to do whatever kind of work may be available. Carpenters with a well-rounded background can switch from residential building to commercial construction to remodeling jobs, depending on demand.

A high school education is desirable, including courses in carpentry, shop, mechanical drawing, and general mathematics. Manual dexterity, eye-hand coordination, physical fitness, and a good sense of balance are important. The ability to solve arithmetic problems quickly and accurately also is helpful.

Employers and apprenticeship committees generally view favorably training and work experience obtained in the armed services and the job corps.

Carpenters may advance to carpentry supervisors or general construction supervisors. Carpenters usually have greater opportunities than most other construction workers to become general construction supervisors because they are exposed to the entire construction process. Some carpenters become independent contractors. To advance, carpenters should be able to estimate the nature and quantity of materials needed to properly complete a job. They also must be able to estimate with accuracy how long a job should take to complete and its cost.

JOB OUTLOOK

Carpenters—the largest group of building trades workers—held about 996,000 jobs in 1996. Four of every five worked for contractors who build, remodel, or repair buildings and other structures. Most of the remainder worked for manufacturing firms, government agencies, wholesale and retail establishments, and schools. About four of every ten were self-employed.

Carpenters are employed throughout the country in almost every community.

Job opportunities for carpenters are expected to be plentiful through the year 2006, due primarily to extensive replacement needs. Thousands of job openings will become available each year as carpenters transfer to other occupations or leave the labor force. The total number of job openings for carpenters each year usually is greater than for other craft occupations because the occupation is large and turnover is high. Since there are no strict training requirements for entry, many people with limited skills take jobs as carpenters but eventually leave the occupation because they find they dislike the work or cannot obtain steady employment.

Increased demand for carpenters will create additional job openings. Employment is expected to increase more slowly than the average for all occupations through the year 2006.

Construction activity should increase slowly in response to demand for new housing and commercial and industrial plants and the need to renovate and modernize existing structures. Opportunities for frame carpenters will be particularly good. The demand for carpenters will be offset somewhat by expected productivity gains resulting from the increasing use of prefabricated components—such as prehung doors and windows and prefabricated wall panels and stairs—that can be installed much more quickly. Prefabricated walls, partitions, and stairs can be quickly lifted into place in one operation; beams, and in some cases entire roof assemblies, can be lifted into place using a crane. As prefabricated components become more standardized, their use will continue to increase. In addition, stronger adhesives that reduce the time needed to join materials and lightweight cordless pneumatic and combustion tools—such as nailers and drills, as well as sanders with electronic speed controls—will make carpenters more efficient and will also reduce fatigue.

Although employment of carpenters is expected to grow over the long run, people entering the occupation should expect to experience periods of unemployment. This results from the short-term nature of many construction projects and the cyclical nature of the construction industry. Building activity depends on many factors—interest rates, availability of mortgage funds, government spending, and business investment—that vary with the state of the economy. During economic downturns, the number of job openings for carpenters declines.

Carpenters with all-around skills will have better business opportunities than those who can do only relatively simple, routine tasks. Therefore, carpenters should always be up-to-date on the new and improved tools, equipment, techniques, and materials so that they may vastly increase their versatility and employability.

Job opportunities for carpenters also vary by geographic area. Construction activity parallels the movement of people and businesses and reflects differences in local economic conditions. Because of this, the number of job opportunities and apprenticeship opportunities in a given year may vary widely from area to area.

SALARIES

Median weekly earnings of carpenters, excluding the self-employed, were $476 in 1996. The middle 50 percent earned between $345 and $660 per week. Weekly earnings for the top 10 percent of all carpenters were more than $874, while the lowest 10 percent earned less than $267.

Earnings may be reduced on occasion because carpenters lose work time in bad weather and during recessions when jobs are unavailable.

Many carpenters are members of the United Brotherhood of Carpenters and Joiners of America.

RELATED FIELDS

Carpenters are skilled construction workers. Workers in other skilled construction occupations include bricklayers, concrete masons, electricians, pipe fitters, plasterers, plumbers, stonemasons, and terrazzo workers.

INTERVIEW
Lawney Schultz
Carpenter

Lawney Schultz, now retired, was a journeyman carpenter for 25 years, working in commercial and residential carpentry, with his most recent 8 years in the Puyallup School District #3 in Puyallup, Washington, where he was responsible for building maintenance. He was in the field from 1971 through 1995. He attended a preapprenticeship program for three months, and a carpentry apprenticeship program for four years.

How Lawney Schultz Got Started

"Working in retail sales and buying for major chain stores for 10 years, I became disillusioned with the management race, to meet or beat 'last year's figures.' A friend of mine told me about a laborer position to help make ends meet while I was looking for another job. I found I enjoyed working with my hands. I asked the contractor what I needed to do to get into an apprenticeship program. I was interested in the 'building aspect' versus just labor. I thought it would be a way to make my mark on the world and leave something behind.

"I have assisted in building hospitals, supermalls, and restaurants. It is wonderful to drive past a building and know I had a hand in creating such monuments to society. I also contracted building two homes on my own with one assistant and have restored turn-of-the-century homes.

"I tried 'in-between jobs' to get hired by a school district or a hospital as a building maintenance carpenter. When an advertisement came out in the local paper for my most recent position, I applied."

What the Job's Really Like

"Commercial carpentry is hustle, hustle, hustle. It is fast paced due to time schedules and deadlines. Often there are penalty clauses connected to the contract, which will be invoked if work is not completed by a certain date. The work is enjoyable, but it can be frustrating because of the elements, bad weather, or not having materials available when they are needed. Days go by quickly, most of the time.

"The carpenter on a construction site is responsible for the organization of all the other trade workers. The superintendent on a construction job is normally a seasoned or 'master carpenter' who organizes all of the other trades (i.e., electricians, plumbers, HVAC, and iron workers) and brings them in on a job as needed.

"Construction is nothing at all like white-collar work; there are no magical three strikes, you're out. If your foreman or superintendent dislikes your attitude, your work, or anything about you at all—they can discharge you.

"You work with a crew, yet you are usually paired with only one other person. This is necessary for safety reasons. So, although there may be a fair number of workers on a job site, you work relatively by yourself.

"Most of my work in the early years was all outdoor work—which was enjoyable most of the time, but during extreme weather conditions it was the most miserable work. However, when you're done with the day or the job, the feeling of elation overshadows all of those wet, stormy days. You feel healthy, strong, and alive in a way the average indoor worker never will.

"There is a lot of concrete work involved with carpentry for the building of forms for walls, foundations, panels, and columns. There is a lot of high and overhead work, as well as below-surface ground work. You start at the bottom and work your way up, literally. The first-year apprentices are involved in a lot of demolition and form-stripping work—no building whatsoever. They must learn to tear it down before they can put it together.

"There are times when you are working in a rural area building warehouses or a major store, and you can see hawks, eagles, coyotes, rare birds, and curious deer interested in what you are doing in their world.

"Being able to take a nonmanufactured object and turn it into a unique building or window dressing is a learning experience and very creative. Blueprints don't tell you how to build a building; they only show you what it looks like when it's done. You still have to put the right things together in the right way to make it come out, not only according to the prints, but also so it's safe for public use.

"Learning to work with compound angles, pitches, and rises requires very accurate mathematical calculation and skill to make the tools make those kinds of cuts. Even though there is a lot of wood around, you still need accuracy in your measurements and cuts to eliminate waste. Waste costs the employer money and can eventually cost you your job. An example of this is: Some building panels (four feet by eight feet) can run from $5 to $100 per panel. One of the golden rules of carpentry is: 'Measure everything twice (at least) but cut only once.' The reason is that if you cut it too long, it's not a problem, but if you cut it too short, it's a complete waste of wood and/or materials.

"Typical tools in this trade are wide and varied: everything from pencils to stationary power tools, mobile power tools, and handheld power tools, as well as nonpower tools. Because many of the tools belong to you, you must learn how to care for the equipment. And learning how to use tools properly is as important as knowing what tool to use for what job. Tools can be very expensive. A quality tool is worth its weight in gold, though, because it will last you a lifetime if properly taken care of. It is an unwritten rule that if someone borrows one of your tools and breaks it, the borrower will replace the tool within 24 hours. Having enough tools of your own to do the job is very important and can mean the difference between being hired and not being hired.

"The reason I wanted to work for a large school district is that you are not stuck in one place. I spent many years in the trade, and always seeing a new job site is part of the excitement of this job. Also, you meet new and different people, and that keeps your mind active and stimulated.

"The rewards are when the teachers and principals are grateful to have me there, fixing or building something new for them. On the downside, there are times they don't want you there during school hours at all. They'd want you to make an appointment to do the work after school hours—which is not practical or possible in most cases.

"You never run out of work. Some of the work orders were three years old and others as current as that morning, depending on the priority of the item. Obviously, a leaking roof or another emergency would take precedent.

"As a carpenter for the school district, I did more than an average carpenter would be required to accomplish. Because of the shortage of maintenance people, we would be required to assist some of the other tradespeople in maintenance, such as locksmiths; glaziers; painters; roofers; heating, venting, and air-conditioning (HVAC) workers; boilermakers; electricians; plumbers; and sometimes the ground maintenance workers.

"In the school district I worked for, maintenance personnel all belong to the same union, so there are no conflicts of interest when one is called upon to assist another tradesperson. However, this is not true of all districts; some trades still belong to separate unions.

"My particular job as a carpenter involved a variety of specialties. This even included building office furniture and cabinets for the administration.

"I have learned more through working in many different trades than I would have learned as strictly a commercial carpenter. These include ceramic tile repair and installation, vinyl tile repair, acoustical tile replacement, repairing of gymnasium bleachers (seats and framework) and 20-foot-high extending doors, roof repairs, lock repairs, countertop replacement and repair, and removal and remaking of insulated window panes.

"The life of a carpenter can be difficult on a marriage because of the ups and downs of the general economy, and being out of work is not unusual for the first two years or so. After that, it gets better because employers will begin to ask for you as your work becomes more known in the field.

"As a carpenter, I started out at approximately $16 per hour. Outside commercial work at that time paid approximately $21 per hour. The current pay scale is approximately $24 per hour. There are also a multitude of benefits, such as sick, emergency, and holiday leave, plus a paid vacation of 10 days the first year, and one additional day each year for 10 years. Union workers are required to have money taken out of their pay, which is then held in an account for them to use once per year for a 'paid' vacation (it is $1 per hour and is similar to a 'forced' savings account). Retirement and medical, dental, and optical benefits are also provided. Only union workers receive these benefits, though."

Expert Advice

"To be a carpenter, you must love the outdoors, be physically fit, have good stamina, be healthy, and possess a basic mechanical aptitude, an ability to read blueprints, and good math skills.

"You must participate in an apprenticeship program if you are serious about learning the trade. You need classroom work as well as on-the-job-training to fully understand all that you will be required to know. The more experience you get in all facets in wood, concrete, and metalwork, the more steadily you will be employed because your reputation will precede you.

"Many times a superintendent or foreman will judge you by the number of tools you have in your toolbox. This tells people how much experience you have and/or your concern for doing a quality job."

FOR MORE INFORMATION

For information about carpentry apprenticeships or other work opportunities in this trade, contact local carpentry contractors, locals of the union, local joint union-contractor apprenticeship committees, or the nearest office of the state employment service or state apprenticeship agency.

For general information about carpentry, contact:

Associated Builders and Contractors
1300 North 17th Street
Rosslyn, VA 22209

Associated General Contractors of America, Inc.
1957 E Street NW
Washington, D.C. 20006-5199

Home Builders Institute
National Association of Home Builders
1201 15th Street NW
Washington, D.C. 20005

United Brotherhood of Carpenters and Joiners of America
101 Constitution Avenue NW
Washington, D.C. 20001

CHAPTER 7 Roofers

EDUCATION
H.S. Preferred

$$$ SALARY
$11,000 to $33,000

OVERVIEW

A leaky roof can damage ceilings, walls, and furnishings. To protect buildings and their contents from water damage, roofers repair and install roofs of tar or asphalt and gravel, rubber, thermoplastic, and metal; and shingles made of asphalt, slate, fiberglass, wood, tile, or other material. Repair and reroofing—replacing old roofs on existing buildings—provide many work opportunities for these workers. Roofers also may waterproof foundation walls and floors.

There are two types of roofs: flat and pitched (sloped). Most commercial, industrial, and apartment buildings have flat or slightly sloping roofs. Most houses have pitched roofs. Some roofers work on both types; others specialize.

Most flat roofs are covered with several layers of materials. Roofers first put a layer of insulation on the roof deck. Over the insulation, they spread a coat of molten bitumen, a tarlike substance. Next, they install partially overlapping layers of roofing felt—a fabric saturated in bitumen—over the insulation surface and use a mop to spread hot bitumen over it and under the next layer. This seals the seams and makes the surface watertight. Roofers repeat these steps to build up the desired number of layers, called "plies." The top layer either is glazed to make a smooth finish or has gravel embedded in the hot bitumen for a rough surface.

An increasing number of flat roofs are covered with a single-ply membrane of waterproof rubber or thermoplastic compounds. Roofers roll these sheets over the roof's insulation and seal the seams. Adhesives, mechanical fasteners, or stone ballast hold the sheets in place. The building must be of sufficient strength to hold the ballast.

Most residential roofs are covered with shingles. To apply shingles, roofers first lay, cut, and tack three-foot strips of roofing felt lengthwise over the entire roof. Then, starting from the bottom edge, they nail overlapping rows of shingles to the roof. Workers measure and cut the felt and shingles to fit intersecting roofs, and to fit around vent pipes and chimneys. Wherever two roof surfaces intersect or shingles reach a vent pipe or chimney, roofers cement or nail "flashing," strips of metal or shingle, over the joints to make them watertight. Finally, roofers cover exposed nailheads with roofing cement or caulking to prevent water leakage.

Some roofers also waterproof and dampproof masonry and concrete walls and floors. To prepare surfaces for waterproofing, they hammer and chisel away rough spots or remove them with a rubbing brick before applying a coat of liquid waterproofing compound. They also may paint or spray surfaces with a waterproofing material or attach waterproofing membrane to surfaces. When dampproofing, they usually spray a bitumen-based coating on interior or exterior surfaces.

Roofers' work is strenuous. It involves heavy lifting, as well as climbing, bending, and kneeling. Roofers risk injuries from slips or falls from scaffolds, ladders, or roofs, and burns from hot bitumen. In fact, of all construction industries, the roofing industry has the highest accident rate. Roofers work outdoors in all types of weather, particularly when making repairs. Roofs are extremely hot during the summer.

TRAINING

Most roofers acquire their skills informally by working as helpers for experienced roofers. They start by carrying equipment and material and erecting scaffolds and hoists. Within two or three months, they are taught to measure, cut, and fit roofing materials and then to lay asphalt or fiberglass shingles. Because some roofing materials are used infrequently, it can take several years to get experience working on all the various types of roofing applications.

Some roofers train through three-year apprenticeship programs administered by local union-management committees representing roofing contractors and locals of the United Union of Roofers, Waterproofers, and Allied Workers. The apprenticeship program generally consists of a minimum of 1,400 hours of on-the-job training annually, plus 144 hours of classroom instruction a year in subjects such as tools and their use, arithmetic, and safety.

On-the-job training for apprentices is similar to that for helpers, except that the apprenticeship program is more structured. Apprentices also learn to dampproof and waterproof walls.

Good physical condition and good balance are essential for roofers. A high school education or its equivalent is helpful, as are courses in mechanical drawing and basic mathematics. Most apprentices are at least 18 years old.

Roofers may advance to supervisor or estimator for a roofing contractor or become contractors themselves.

JOB OUTLOOK

Roofers hold about 126,000 jobs. Almost all wage and salary roofers work for roofing contractors. Nearly one-third of all roofers are self-employed. Many self-employed roofers specialize in residential work.

Jobs for roofers should be plentiful through the year 2005, primarily because of the need to replace workers who transfer to other occupations or who leave the labor force.

Turnover is high; roofing work is hot, strenuous, and dirty, and a significant number of workers treat roofing as a temporary job until something better comes along. Some roofers leave the occupation to go into other construction trades.

Employment of roofers is expected to increase about as fast as the average for all occupations through the year 2005. Roofs deteriorate faster than most other parts of buildings and periodically need to be repaired or replaced. About 75 percent of roofing work is repair and reroofing, a higher proportion than in most other construction work. As a result, demand for roofers is less susceptible to downturns in the economy than some of the other construction trades.

In addition to repair and reroofing work on the growing stock of buildings, new construction of industrial, commercial, and residential buildings will add to the demand for roofers. However, many innovations and advances in materials, techniques, and tools have made roofers more productive and will restrict the growth of employment at least to some extent. Jobs should be easiest to find during spring and summer, when most roofing is done.

SALARIES

Median weekly earnings for roofers working full time is about $371. The middle 50 percent earn between $278 and $498 a week. The top 10 percent earn more than $630 weekly, and the lowest 10 percent earn less than $219 a week.

According to the *Engineering News Record,* average hourly earnings, including benefits, for union roofers are about $25. Wages ranged from a low of $13.90 in Denver to a high of $38.58 in New York City. Apprentices generally start at about 40 percent of the rate paid to experienced roofers and receive periodic raises as they acquire the skills of the trade. Earnings for roofers are reduced on occasion because poor weather often limits the time they can work.

RELATED FIELDS

Roofers use shingles, bitumen and gravel, single-ply plastic or rubber sheets, or other materials to waterproof building surfaces.

Workers in other occupations who cover surfaces with special materials for protection and decoration include carpenters, concrete masons, drywall installers, floor covering installers, plasterers, terrazzo workers, and tile setters.

INTERVIEW
Kris Irr
Roofing Contractor

Kris Irr is vice president of a roofing contracting firm in Boca Raton, Florida. He earned his B.S.E.E. from Florida Atlantic University and has a state-certified contractor's license. He has been in the field since 1992.

How Kris Irr Got Started

"I was flat broke, and a contractor friend let me sell for him. I sold new roofs for a six-building condo, but the pay was poor, so I decided to become a contractor myself. I was able to save a little money from the first job and then bought the equipment I needed to get started."

What the Job's Really Like

"I wake up at 5 A.M., stop at a couple of supply houses for materials or repair parts, then I'm in the office by 7. I work on proposals for new work, estimating costs, and organizing the work list for the crews.

"At 8 A.M. we load the trucks, and I assign the work to the crews. I then pull permits or meet with clients and engineers. I visit the sites, check on the progress of the work, and handle calls and emergencies on my cell phone.

"At the end of the day, about 4 or 5 P.M., I receive the crews back in my office and assess the performance. Then I prepare for the next day's needs and complete book work—the accounting and billing. I might have more estimates to prepare and correspondence to take care of. Sometimes vehicles or equipment need repair.

"I'm home by 9 P.M., asleep by 10:30, and get up the next morning and repeat the same. I also work on the weekends, a minimum of four hours, except when we're behind, and then it could be twelve hours both days. Any time off I have I spend with my children.

"What I enjoy most with this work is figuring out solutions to the problems encountered on construction sites and also the profit I earn. Of course, I don't enjoy losing any profit. I earn enough to pay my bills, but someone just starting out could expect to starve."

Expert Advice

"You must be incredibly self-motivated to succeed in this business. Do your homework; learn from credible sources. Provide quality work at a fair price, and don't take low-bid work."

FOR MORE INFORMATION

For information about roofing apprenticeships or work opportunities in this trade, contact local roofing contractors; a local of the roofers union; a local joint union-management apprenticeship committee; or the nearest office of the state employment service or state apprenticeship agency.

For information about the work of roofers, contact:

National Roofing Contractors Association
10255 West Higgins Road
Rosemont, IL 60018

United Union of Roofers, Waterproofers, and Allied Workers
1125 17th Street NW
Washington, D.C. 20036

CHAPTER 8 Plumbers and Pipe Fitters

EDUCATION
H.S. Preferred

$$$ SALARY
$16,000 to $54,000

OVERVIEW

Most people are familiar with plumbers who come to your home to unclog a drain or install an appliance. In addition to these activities, however, plumbers and pipe fitters install, maintain, and repair many different types of pipe systems. For example, some systems move water to a municipal water treatment plant, and then to residential, commercial, and public buildings. Others dispose of waste. Some bring in gas for stoves and furnaces. Others supply air-conditioning. Pipe systems in power plants carry the steam that powers huge turbines. Pipes also are used in manufacturing plants to move material through the production process.

Although plumbing and pipe fitting sometimes are considered a single trade, workers generally specialize in one or the other. Plumbers install and repair the water, waste disposal, drainage, and gas systems in homes and commercial and industrial buildings. They also install plumbing fixtures—bathtubs, showers, sinks, and toilets—and appliances such as dishwashers and water heaters.

Pipe fitters install and repair both high- and low-pressure pipe systems that are used in manufacturing, in the generation of electricity, and in heating and cooling buildings. They also install automatic controls that are increasingly being used to regulate these systems. Some pipe fitters specialize in only one type of system. Steamfitters, for example, install pipe systems that move liquids or gases under high pressure. Sprinkler fitters install automatic fire sprinkler systems in buildings.

Plumbers and pipe fitters use many different materials and construction techniques, depending on the type of project. Residential water systems, for example, use copper, steel, and increasingly, plastic pipe that can be handled and installed by one or two workers. Municipal sewerage systems, on the other hand, are made of large cast-iron pipes; installation normally requires crews of pipe fitters. Despite these differences, all plumbers and pipe fitters must be able to follow building plans or blueprints and instructions from supervisors, lay out the job, and work efficiently with the materials and tools of the trade.

When construction plumbers install piping in a house, for example, they work from blueprints or drawings that show the planned location of pipes, plumbing fixtures, and appliances. They lay out the job to fit the piping into the structure of the house with the least waste of material and within the confines of the structure. They measure and mark areas where pipes will be installed and connected. They check for obstructions, such as electrical wiring, and, if necessary, plan the pipe installation around the problem.

Sometimes plumbers have to cut holes in walls, ceilings, and floors of a house. For some systems, they may have to hang steel supports from ceiling joists to hold the pipe in place. To assemble the system, plumbers cut and bend lengths of pipe using saws, pipe cutters, and pipe-bending machines. They connect lengths of pipe with fittings; the method depends on the type of pipe used. For plastic pipe, plumbers connect the sections and fittings with adhesives. For copper pipe, they slide fittings over the end of the pipe and solder the fitting in place with a torch.

After the piping is in place in the house, plumbers install the fixtures and appliances and connect the system to the outside water or sewer lines. Using pressure gauges, they check the system to ensure that the plumbing works properly.

Because plumbers and pipe fitters must frequently lift heavy pipes, stand for long periods, and sometimes work in uncomfortable or cramped positions, they need physical strength as well as stamina. They may have to work outdoors in inclement weather. They also are subject to falls from ladders, cuts from sharp tools, and burns from hot pipes or soldering equipment.

Plumbers and pipe fitters engaged in construction generally work a standard 40-hour week; those involved in maintaining pipe systems, including those who provide maintenance services under contract, may have to work evening or weekend shifts, as well as be on call. These maintenance workers may spend quite a bit of time traveling to and from work sites.

TRAINING

Virtually all plumbers undergo some type of apprenticeship training. Many programs are administered by local union-management committees made up of members of the United Association of Journeymen and Apprentices of the Plumbing and Pipe Fitting Industry of the United States and Canada, and local employers who are members of either the Mechanical Contractors Association of America, the National Association of Plumbing-Heating-Cooling Contractors, or the National Fire Sprinkler Association, Inc.

Nonunion training and apprenticeship programs are administered by local chapters of the Associated Builders and Contractors, the National Association of Plumbing-Heating-Cooling Contractors, the American Fire Sprinkler Association Inc., and the Home Builders Institute of the National Association of Home Builders. (Addresses are provided at the end of the chapter.)

Apprenticeships—both union and nonunion—consist of four to five years of on-the-job training, in addition to at least 144 hours annually of related classroom instruction. Classroom subjects include drafting and blueprint reading, mathematics, applied physics and chemistry, safety, and local plumbing codes and regulations. On the job, apprentices first learn basic skills such as identifying grades and types of pipe, the use of the tools of the trade, and the safe unloading of materials. As apprentices gain experience, they learn how to work with various types of pipe and install different piping systems and plumbing fixtures. Apprenticeship gives trainees a thorough knowledge of all aspects of the trade. Although most plumbers are trained through apprenticeship, some still learn their skills informally on the job.

Applicants for union or nonunion apprentice jobs must be at least 18 years old and in good physical condition. Apprenticeship committees may require applicants to have a high school diploma or its equivalent. Armed forces training in plumbing and pipe fitting is considered very good preparation. In fact, persons with this background may be given credit for experience when entering a civilian apprenticeship program. Secondary or postsecondary courses in shop, plumbing, general mathematics, drafting, blueprint reading, and physics also are good preparation.

Although there are no uniform national licensing requirements, most communities require plumbers to be licensed. Licensing requirements vary from area to area, but most localities require workers to pass an examination that tests their knowledge of the trade and of local plumbing codes.

Some plumbers and pipe fitters may become supervisors for mechanical and plumbing contractors. Others go into business for themselves.

JOB OUTLOOK

Plumbers and pipe fitters hold about 400,000 jobs nationwide. About two-thirds work for mechanical and plumbing contractors engaged in new construction, repair, modernization, or maintenance work. Others do maintenance work for a variety of industrial, commercial, and government employers. For example, pipe fitters are employed as maintenance personnel in the petroleum and chemical industries, where manufacturing operations require the moving of liquids and gases through pipes.

One of every five plumbers and pipe fitters is self-employed.

Jobs for plumbers and pipe fitters are distributed across the country in about the same proportion as the general population.

Job opportunities for skilled plumbers and pipe fitters are expected to be good as the growth in demand outpaces the supply of workers trained in this craft. Employment of plumbers and pipe fitters is expected to grow more slowly than the average for all occupations through the year 2005. However, the pool of young workers available to enter training programs will also be increasing slowly, and many in that group are reluctant to seek training for jobs that may be strenuous and have uncomfortable working conditions.

In addition, several thousand positions will become available each year from the need to replace experienced workers who leave the occupation.

Construction activity—residential, industrial, and commercial—is expected to grow slowly over the next decade. Demand for plumbers will stem from building renovation, including the increasing installation of sprinkler systems; repair and maintenance of existing residential systems; and maintenance activities for places that have extensive systems of pipes, such as power plants, water and wastewater treatment plants, pipelines, office buildings, and factories. However, the growing use of plastic pipe and fittings, which are much easier to use; more efficient sprinkler systems; and other technologies will mean that employment will not grow as fast as it has in past years.

Traditionally, many organizations with extensive pipe systems have employed their own plumbers or pipe fitters to maintain their equipment and keep everything running smoothly. But, in order to reduce their labor costs, many of these firms no longer employ a full-time in-house plumber or pipe fitter. Instead, when they need one, they rely on workers provided by plumbing and pipe fitting contractors under service contracts.

All construction projects provide only temporary employment, so when a project ends, plumbers and pipe fitters working on it may experience short bouts of unemployment. Because construction activity varies from area to area, job openings, as well as apprenticeship opportunities, fluctuate with local economic conditions. However, employment of plumbers and pipe fitters generally is less sensitive to changes in economic conditions than some of the other construction trades. Even when construction activity declines, maintenance, rehabilitation, and replacement of existing piping systems, as well as the growing installation of fire sprinkler systems, provide many jobs for plumbers and pipe fitters.

SALARIES

Median weekly earnings for plumbers and pipe fitters who were not self-employed were $591 in 1996. The middle 50 percent earned between $413 and $812 weekly. The lowest 10 percent earned less than $312; the highest 10 percent earned more than $1,047 a week.

In 1996, the latest figures available, the median hourly wage rate for maintenance pipe fitters in 160 metropolitan areas was about $21.46. The middle 50 percent earned between $19.20 and $21.65 an hour.

In general, wage rates tend to be higher in the Midwest and West than in the Northeast and South.

Apprentices usually begin at about 50 percent of the wage rate paid to experienced plumbers or pipe fitters. This pay increases periodically as they improve their skills. After an initial waiting period, apprentices receive the same benefits as experienced plumbers and pipe fitters.

Many plumbers and pipe fitters are members of the United Association of Journeymen and Apprentices of the Plumbing and Pipe Fitting Industry of the United States and Canada.

RELATED FIELDS

Other occupations in which workers install and repair mechanical systems in buildings are boilermakers, stationary engineers, electricians, elevator installers, industrial machinery repairers, millwrights, sheet-metal workers, and heating, air-conditioning, and refrigeration mechanics.

INTERVIEW
Joseph C. Elliott
Steam and Sprinkler Fitter

Joseph Elliott has been working with the Port Authority of New York and New Jersey since 1985. He was hired as a laborer and learned the trade on the job.

How Joseph C. Elliott Got Started

"I have worked for the Port Authority for 13 years, though I have been a steam and sprinkler fitter for 8 years. I was hired as a laborer and learned the trade on the job. In the summer of 1989 I took the test to become a fitter and was promoted in February of 1990.

"When the Port Authority hired me, all I had was a high school diploma. I returned to school in September of 1993, and in December of 1998 I earned an Associate in Applied Science degree in business administration from Kingsborough Community College. I plan to continue on for my bachelor's degree starting next year.

"What attracted me the most to the job was the fact that most of the work I do is involved with fire prevention systems, specifically the repair and maintenance of fire sprinklers and fire standpipe and their various subsystems.

"The Port Authority has been, essentially, the only real job I've ever had. I was hired when I was 18. Before then I worked part time after school and on weekends at various odd jobs.

"I found out about working for the Port Authority when I went back to my high school to pick up my diploma. I bumped into a teacher who told me that the Port Authority was hiring laborers to work in its maintenance department. All that was required was that I have a high school diploma from a vocational high school, which I just so happened to have in my hands. That same day, I went to their main office and filled out an application. I was later called to take a written test, a hands-on practical, and then a physical examination. In November of 1985 I was hired."

What the Job's Really Like

"I work in the maintenance department, and so the focus of my work is on the care and repair of all of the equipment related to my field—valves, pipes, air compressors, pumps, underground water mains, fire hydrants, fire extinguishers, and alarm valves, just to name a few.

"The job is structured so that for part of the week the mechanics inspect all of the equipment, looking for problems and deficiencies, and the other part of the week is spent making repairs. That is the way the system is supposed to work, but it sometimes doesn't. Very often, something that was checked the day before, and was operating just fine, decides to break on you the next and usually at some ungodly time of day—usually as you're just heading out the door to go home!

"I work a standard 40-hour workweek. My day begins at 6:30 A.M. and ends at 3 in the afternoon. The overall pace of the job is one of two extremes: extremely busy or extremely mundane.

"For the most part, it can be interesting when you're trying to troubleshoot a problem with a piece of equipment. The routine inspections can become boring, especially if all the equipment is in good working order. Then you can start to get lazy and start to overlook things. So, it's important to keep on your toes.

"The downside to all of this, though, is that most of the time I have to work out in the weather, which is really nice in the spring and fall. However, I've had to repair 24-inch water mains 10 feet underground in the middle of January or work on remote fire sprinkler pipes 30 feet in the air in a warehouse constructed of aluminum and steel, in the dog days of a humid New York City summer. In all honesty, if I could pass on it I would, but that is what I'm paid to do.

"Salaries on my job are structured with annual step raises so that every year a person starting on the job could expect a raise until you reach what is called 'top step.' This can take between five and seven years, depending on your job title. The basic rule is that it takes a laborer about seven years and a mechanic, of any trade, about five years to reach top step. After that, your raises are determined by what is negotiated in the union contract. Right now I am a top-step mechanic, and I have an annual base salary of $52,000. That does not include overtime or any other special pay. Someone starting out could expect to earn about $30,000 to $35,000 a year."

Expert Advice

"If you do plan to go into this line of work, my best advice is to go to a trade school before going to work in the field. This way you'll learn all the theory related to the trade in an organized fashion.

"It's also important to note that when you start out on your first job, probably with a privately owned contractor, it will be as a laborer, not as a mechanic. This applies to everybody; you have to start out at the bottom and prove yourself.

"Most important, you should enjoy doing physical labor, and that means getting your hands dirty. If you can't stand the idea, find another vocation; you'll be a lot happier."

INTERVIEW
Max King Jr.
Master Plumber

Max King Jr. works for Local #357 in Kalamazoo, Michigan. He is a high school graduate and studied for two years at a community college where he took classes in plumbing, welding, and math. He has been working in the field since 1985.

How Max King Jr. Got Started

"My dad was a design engineer, and he taught math at night at the local college. He was also a do-it-yourselfer, and he did his own plumbing, electrical work, and building. All my brothers and I worked with him, starting when we were just little tykes, so it was not surprising that one brother ended up owning his own construction company, another ended up a construction foreman, and I ended up in plumbing. I watched and kept my mouth shut, and that was how I learned.

"I started as a laborer. Then one day I was helping out a friend in Detroit who had all his copper pipe stolen. His friend had a boss who saw me doing all the plumbing work and liked what he saw.

"Getting my current job was a fluke. I went to a convenience store and mentioned to the checkout lady that I was a plumber. The man behind me in line spoke up and asked me if I was with the local union. I said, no, but that I would like to be. He called the union; they were looking for qualified plumbers, and they took me in."

What the Job's Really Like

"My duties cover all aspects of the plumbing field. Some days the work is very heavy. I can work with eight-inch cast iron, or six-inch steel pipe, 21 feet long. It involves sawing, measuring, and welding the materials together. It can be extremely hard on the body, especially the back. It makes for long, strenuous days.

"Other days I can be working with half-inch or three-quarter-inch copper pipe or PVC (plastic pipe), which are both very light and very easy to work with, making for a less strenuous day.

"I like it most when I'm running the site as a foreman. I least like service work. Service work is when you go out to houses or businesses and are working with used pipe. The other downside is having to work with replacing or relocating used sewer pipe."

Expert Advice

"First, get four years of a college education and push a pencil. For training, get anything you can find on your own, through community schools, or offered through your place of employment.

"You need to have good mechanical and math skills, as well as a strong back.

"To get started, watch the paper for Plumber's Helper Wanted ads, and keep your mouth shut and your eyes open. Learn by seeing before doing. Try not to ask more than three to four questions a day. But when doing, don't be afraid to ask questions rather than make a stupid, expensive mistake."

FOR MORE INFORMATION

For information about apprenticeships or work opportunities in plumbing and pipe fitting, contact local plumbing, heating, and air-conditioning contractors; a local or state chapter of the National Association of Plumbing-Heating-Cooling Contractors; a local chapter of the Mechanical Contractors Association of America, Inc.; a local of the United Association of Journeymen and Apprentices of the Plumbing and Pipe Fitting Industry of the United States and Canada; or the nearest office of the state employment service or state apprenticeship agency. This information is also available from:

Home Builders Institute
National Association of Home Builders
1201 15th Street NW
Washington, D.C. 20005

For general information about the work of plumbers, pipe fitters, and sprinkler fitters, contact:

American Fire Sprinkler Association, Inc.
12959 Jupiter Road, Suite 142
Dallas, TX 75238-3200

Associated Builders and Contractors
1300 North 17th Street
Rosslyn, VA 22209

Mechanical Contractors Association of America
1385 Piccard Drive
Rockville, MD 20850

National Association of Plumbing-Heating-Cooling
   Contractors
180 South Washington Street
P.O. Box 6808
Falls Church, VA 22046

National Fire Sprinkler Association
P.O. Box 1000
Patterson, NY 12563

United Association of Journeymen and Apprentices of the
   Plumbing and Pipe Fitting Industry of the United States
   and Canada
901 Massachusetts Avenue NW
Washington, D.C. 20001

Heating, Air-Conditioning, and Refrigeration Technicians

OVERVIEW

What would people living in Chicago do without heating, those in Miami do without air-conditioning, or blood banks in all parts of the country do without refrigeration? Heating and air-conditioning systems control the temperature, humidity, and total air quality in residential, commercial, industrial, and other buildings. Refrigeration systems make it possible to store and transport food, medicine, and other perishable items. Heating, air-conditioning, and refrigeration technicians install, maintain, and repair such systems.

Heating, air-conditioning, and refrigeration systems consist of many mechanical, electrical, and electronic components, including motors, compressors, pumps, fans, ducts, pipes, thermostats, and switches. In central heating systems, for example, a furnace heats air that is distributed throughout the building via a system of metal or fiberglass ducts. Technicians must be able to maintain, diagnose, and correct problems throughout the entire system. To do this, they may adjust system controls to recommended settings and test the performance of the entire system using special tools and test equipment.

Although they are trained to do both, technicians generally specialize in either installation or maintenance and repair. Some further specialize in one type of equipment—for example, oil burners, solar panels, or commercial refrigerators. Technicians may work for large or small contracting companies or directly for a manufacturer or wholesaler. Those working for smaller operations tend to do both installation and servicing, and work with heating, cooling, and refrigeration equipment.

Furnace installers, also called heating equipment technicians, follow blueprints or other specifications to install oil, gas, electric, solid-fuel, and multiple-fuel heating systems. After putting the equipment in place, they install fuel and water supply lines, air ducts and vents, pumps, and other components. They may connect electrical wiring and controls and check the unit for proper operation. To ensure the proper functioning of the system, furnace installers often use combustion test equipment such as carbon dioxide and oxygen testers.

After a furnace has been installed, technicians often perform routine maintenance and repair work in order to keep the system operating efficiently. During the fall and winter, for example, when the system is used most, they service and adjust burners and blowers. If the system is not operating properly, they check the thermostat, burner nozzles, controls, or other parts in order to diagnose and then correct the problem. During the summer, when the heating system is not being used, technicians do maintenance work, such as replacing filters and vacuum-cleaning vents, ducts, and other parts of the system that may accumulate dust and impurities during the operating season.

Air-conditioning and refrigeration technicians install and service central air-conditioning systems and a variety of refrigeration equipment. Technicians follow blueprints, design specifications, and manufacturers' instructions to install motors, compressors, condensing units, evaporators, piping, and other components. They connect this equipment to the ductwork, refrigerant lines, and electrical power source. After making the connections, they charge the system with refrigerant, check it for proper operation, and program control systems.

When air-conditioning and refrigeration equipment break down, technicians diagnose the problem and make repairs. To do this, they may test parts such as compressors, relays, and thermostats. During the winter, air-conditioning technicians inspect the systems and do required maintenance, such as over-hauling compressors.

When servicing equipment, heating, air-conditioning, and refrigeration technicians must use care to conserve, recover, and recycle chlorofluorocarbon (CFC) and hydrochlorofluorocarbon (HCFC) refrigerants used in air-conditioning and refrigeration systems. The release of CFCs and HCFCs contributes to the depletion of the stratospheric ozone layer, which protects plant and animal life from ultraviolet radiation. Technicians conserve the refrigerant by making sure that there are no leaks in the system; they recover it by venting the refrigerant into proper cylinders; and they recycle it for reuse with special filter-dryers.

Heating, air-conditioning, and refrigeration technicians use a variety of tools, including hammers, wrenches, metal snips, electric drills, pipe cutters and benders, measurement gauges, and acetylene torches, to work with refrigerant lines and air ducts. They use voltmeters, thermometers, pressure gauges, manometers, and other testing devices to check air flow, refrigerant pressure, electrical circuits, burners, and other components.

Cooling and heating systems sometimes are installed or repaired by other craft workers. For example, on a large air-conditioning installation job, especially where workers are covered by union contracts, ductwork might be done by sheet-metal workers; electrical work by electricians; and installation of piping, condensers, and other components by plumbers and pipe fitters. Room air conditioners and household refrigerators usually are serviced by home appliance repairers.

Heating, air-conditioning, and refrigeration technicians work in homes, supermarkets, hospitals, office buildings, factories—anywhere there is climate control equipment. They may be assigned to specific job sites at the beginning of each day, or if they are making service calls, they may be dispatched to jobs by radio or telephone.

Technicians may work outside in cold or hot weather or in buildings that are uncomfortable because the air-conditioning or heating equipment is broken. In addition, technicians often work in awkward or cramped positions and sometimes are required to work in high places. Hazards include electrical shock, burns, muscle strains, and other injuries from handling heavy equipment. Appropriate safety equipment is necessary when handling refrigerants, since contact can cause skin damage, frostbite, or blindness. Inhalation of refrigerants when working in confined spaces is also a possible hazard, and may cause asphyxiation.

Technicians usually work a 40-hour week, but during peak seasons they often work overtime or irregular hours. Maintenance workers, including those who provide maintenance services under contract, often work evening or weekend shifts, and are on call. Most employers try to provide a full workweek the year round by doing both installation and maintenance work, and many manufacturers and contractors now provide or even require service contracts. In most shops that service both heating and air-conditioning equipment, employment is very stable throughout the year.

TRAINING

Because of the increasing sophistication of heating, air-conditioning, and refrigeration systems, employers prefer to hire people with technical school or apprenticeship training. A sizable number of technicians, however, still learn the trade informally on the job.

Many secondary and postsecondary technical and trade schools, junior and community colleges, and the armed forces offer six-month to two-year programs in heating, air-conditioning, and refrigeration. Students study theory, design, and equipment construction, as well as electronics. They also learn the basics of installation, maintenance, and repair.

Apprenticeship programs are frequently run by joint committees representing local chapters of the Air-Conditioning Contractors of America, the Mechanical Contractors Association of America, the National Association of Plumbing-Heating-Cooling Contractors, and locals of the Sheet Metal Workers' International Association or the United Association of Journeymen and Apprentices of the Plumbing and Pipe Fitting Industry of the United States and Canada. Other apprenticeship programs are sponsored by local chapters of the

Associated Builders and Contractors and the National Association of Home Builders.

Formal apprenticeship programs generally last three or four years and combine on-the-job training with classroom instruction. Classes include subjects such as the use and care of tools, safety practices, blueprint reading, and air-conditioning theory. Applicants for these programs must have a high school diploma or equivalent.

Those who acquire their skills on the job usually begin by assisting experienced technicians. They may begin performing simple tasks such as carrying materials, insulating refrigerant lines, or cleaning furnaces. In time, they move on to more difficult tasks, such as cutting and soldering pipes and sheet metal and checking electrical and electronic circuits.

Courses in shop math, mechanical drawing, applied physics and chemistry, electronics, blueprint reading, and computer applications provide a good background for people interested in entering this occupation. Some knowledge of plumbing or electrical work is also helpful. A basic understanding of microelectronics is becoming more important because of the increasing use of this technology in solid-state equipment controls. Because technicians frequently deal directly with the public, they should be courteous and tactful, especially when dealing with an aggravated customer. They also should be in good physical condition because they sometimes have to lift and move heavy equipment.

All technicians who purchase or work with refrigerants must be certified so that they know how to handle them properly. To become certified to purchase and handle refrigerants, technicians must pass a written examination specific to the type of work in which they specialize.

The three possible areas of certification are: Type I—servicing small appliances, Type II—high-pressure refrigerants, and Type III—low-pressure refrigerants. Exams are administered by organizations approved by the Environmental Protection Agency, such as trade schools, unions, contractor associations, and building groups. Though no formal training is required for certification, training programs designed to prepare workers for the certification examination, as well as for general skills improvement training, are provided by heating and air-conditioning equipment manufacturers; the Refrigeration Service Engineers Society (RSES); the Air-Conditioning Contractors of America; the Mechanical Contractors Association of America; local chapters of the National Association of Plumbing-Heating-Cooling Contractors; and the United Association of Journeymen and Apprentices of the Plumbing and Pipe Fitting Industry of the United States and Canada. RSES, along with some other organizations, also offers basic self-study courses for individuals with limited experience.

In addition to understanding how systems work, technicians must be knowledgeable about refrigerant products as well as legislation and regulation that govern their use. Advancement usually takes the form of higher wages. Some technicians, however, may advance to positions such as supervisor or service manager. Others may move into areas such as sales and marketing. Those with sufficient money and managerial skill can open their own contracting businesses.

JOB OUTLOOK

Heating, air-conditioning, and refrigeration technicians hold about 233,000 jobs nationwide. More than one-half of these work for cooling and heating contractors. The remainder are employed in a wide variety of industries throughout the country, reflecting a widespread dependence on climate control systems.

Some work for fuel oil dealers, refrigeration and air-conditioning service and repair shops, and schools. Others are employed by the federal government, hospitals, office buildings, and other organizations that operate large air-conditioning, refrigeration, or heating systems.

Approximately one of every eight technicians is self-employed.

Job prospects for highly skilled air-conditioning, heating, and refrigeration technicians are very good, particularly those with technical school or formal apprenticeship training to install, remodel, and service new and existing systems. In addition to job openings created by rapid employment growth, thousands of openings will result from the need to replace workers who transfer to other occupations or leave the labor force.

Employment of heating, air-conditioning, and refrigeration technicians is expected to increase faster than the average for all occupations through the year 2005. As the population and economy grow, so does the demand for new residential, commercial, and industrial climate control systems. Technicians who specialize in installation work may experience periods of unemployment when the level of new construction activity declines, but maintenance and repair work usually remains relatively stable. People and businesses depend on their climate control systems and must keep them in good working order, regardless of economic conditions.

Concern for the environment and energy conservation should continue to prompt the development of new energy-saving heating and air-conditioning systems. An emphasis on better energy management should lead to the replacement of older systems and the installation of newer, more efficient systems in existing homes and buildings. Also, demand for maintenance and service work should increase as businesses and home owners strive to keep systems operating at peak efficiency.

Regulations prohibiting the discharge of CFC and HCFC refrigerants and banning CFC production by the year 2000 also should result in demand for technicians to replace many existing systems, or modify them to use new environmentally safe refrigerants. In addition, the continuing focus on improving indoor air quality should contribute to the growth of jobs for heating, air-conditioning, and refrigeration technicians.

SALARIES

Median weekly earnings of air-conditioning, heating, and refrigeration technicians who work full-time are about $510. The middle 50 percent earn between $375 and $685. The lowest 10 percent earn less than $255 a week, and the top 10 percent earn more than $835 a week.

Apprentices usually begin at about 50 percent of the wage rate paid to experienced workers. As they gain experience and improve their skills, they receive periodic increases until they reach the wage rate of experienced workers.

Heating, air-conditioning, and refrigeration technicians enjoy a variety of employer-sponsored benefits. In addition to typical benefits such as health insurance and pension plans, some employers pay for work-related training and provide uniforms, company vans, and tools.

Nearly one out of every five heating, air-conditioning, and refrigeration technicians is a member of a union. The unions to which the greatest numbers of technicians belong are the Sheet Metal Workers' International Association and the United Association of Journeymen and Apprentices of the Plumbing and Pipe Fitting Industry of the United States and Canada.

RELATED FIELDS

Heating, air-conditioning, and refrigeration technicians work with sheet metal and piping, and repair machinery, such as electrical motors, compressors, and burners. Workers who have similar skills are boilermakers, electrical appliance servicers, electricians, plumbers and pipe fitters, sheet-metal workers, and duct installers.

INTERVIEW
Troy Scott McClure
Owner of McClure Refrigeration, Heating, and Air-Conditioning

Troy Scott McClure owns his own refrigeration, heating, and air-conditioning business in Iowa Park, Texas. He earned his B.S. in education in 1996 from Midwestern State University in Wichita Falls, Texas. He has been in the HVAC field since 1983.

How Troy Scott McClure Got Started

"I had a terrible job at a car dealership, and an opportunity to move on came up in the AC business. Basically, I learned as I went. My boss showed me some things, but his philosophy was that you learn best by actually doing. Now I have attended a few seminars put on by manufacturers of AC equipment.

"Timing is everything in life; things just worked out for me. My father was working with another fireman in 1983 who had an air-conditioning business on the side. He was short of help one day and casually mentioned it around the dinner table at the firehouse where my father worked. The rest is, of course, history.

"After a while I became fed up working for someone else and making him lots of money, so I ventured out on my own."

What the Job's Really Like

"My duties are installing AC equipment, servicing, sending bills, collecting bills, and other boring-type book work. I am the busiest when the weather is either really hot or really cold.

"There is really no such thing as a typical day in my business. This summer has really been hot, so usually I'll get up at daybreak to get any attic work done, then I move on to service work. Usually I eat on the go. My record for service calls in one day is 10, leaving them all running cool. Other days are spent installing equipment. A total system change can take a whole day, and the challenge is to make service calls too.

"It can go from relaxed to swamped in minutes. Usually the fall and spring months are more relaxed, with the busiest times being in the summer.

"The work is interesting at times. Some customers can be a blast, while others can be a pain. But, being self-employed, I am selective in whom I work for. There is a lot of driving between jobs—which can be a drag.

"Summer months I can work 70 hours if my body allows, while in the mild weather I may not work 30 hours a week. Of course, being self-employed is great. The more I work, the more money I make. I don't have the desire to become a huge company, so mainly I do what I can do. Right now my retired fireman father helps me occasionally when I need the extra help. I enjoy meeting new customers and taking care of my old customers' needs. I tend to develop a relationship with my customers and, as strange as that might sound, also with their equipment.

"I like being outside during most of the work, but the downside is there is attic work quite often, and even having to get under houses. Getting under houses is the least desirable of my duties. The most desirable is collecting my money after the job is done. You always worry about getting stiffed on a job.

"The upside is that you can make good money! I charge $36 per hour. The good money is when you can sell equipment, though. Also, on parts you get the markup. You get paid for the labor as well as the markup on the equipment.

"Duct runners start out at the lowest, around $6 an hour. New service techs make about $10 an hour. This area of the country is one of the lowest paying of anywhere, though."

Expert Advice

"Bottom line—this business is not for wimps. If you can't stand the heat, you gotta get out of Dodge. This trade and the others give you some of the few opportunities to become your own boss someday. I worked for my old boss for five years and then went on my own. I learned as I went, basically, but there are trade schools for the profession. The key is to find a good boss that will give you the freedom to be your own person. There are lots of awful people to work for, but I was fortunate to get to work for a great man who helped me on my road to self-employment.

"One more thing: There are lots of bad guys out there in the trades who are very dishonest and try to get as much money from their customers as possible. I despise these guys. I go home every night feeling good about taking care of my people and charging them a fair price."

FOR MORE INFORMATION

For more information about employment and training opportunities in this trade, contact local vocational and technical schools; local heating, air-conditioning, and refrigeration contractors; a local of the unions mentioned; a local joint union-management apprenticeship committee; a local chapter of the

Associated Builders and Contractors; or the nearest office of the state employment service or state apprenticeship agency.

For information on career opportunities and training, write to:

Air Conditioning and Refrigeration Institute
4301 North Fairfax Drive, Suite 425
Arlington, VA 22203

Air-Conditioning Contractors of America
1712 New Hampshire Avenue NW
Washington, D.C. 20009

Associated Builders and Contractors
1300 North 17th Street
Rosslyn, VA 22209

Home Builders Institute
National Association of Home Builders
1201 15th Street NW
Washington, D.C. 20005

Mechanical Contractors Association of America
1385 Piccard Drive
Rockville, MD 20850-4329

National Association of Plumbing-Heating-Cooling
    Contractors
180 South Washington Street
P.O. Box 6808
Falls Church, VA 22046

New England Fuel Institute
P.O. Box 9137
Watertown, MA 02272

Refrigeration Service Engineers Society
1666 Rand Road
Des Plaines, IL 60016-3552

United Association of Journeymen and Apprentices of the
    Plumbing and Pipe Fitting Industry of the United States
    and Canada
901 Massachusetts Avenue NW
Washington, D.C. 20001

CHAPTER 10

Drywall Workers, Lathers, and Plasterers

EDUCATION
H.S. Preferred

$$$ SALARY
$10,000 to $45,000

OVERVIEW

Drywall consists of a thin layer of gypsum sandwiched between two layers of heavy paper. It is used today for walls and ceilings in most buildings because it is both faster and cheaper to install than plaster.

There are two kinds of drywall workers: installers and finishers. Installers, also called applicators, fasten drywall panels to the inside framework of residential houses and other buildings. Finishers, or tapers, prepare these panels for painting by taping and finishing joints and imperfections.

Because drywall panels are manufactured in standard sizes—usually 4 feet by 8 or 12 feet—installers must measure, cut, and fit some pieces around doors and windows. They also saw or cut holes in panels for electrical outlets, air-conditioning units, and plumbing. After making these alterations, installers may glue, nail, or screw the wallboard panels to the wood or metal framework. Because drywall is heavy and cumbersome, a helper generally assists the installer in positioning and securing the panel. A lift is often used when placing ceiling panels.

After the drywall is installed, finishers fill joints between panels with a joint compound. Using the wide, flat tip of a special trowel, they spread the joint compound into and along each side of the joint with brushlike strokes. They immediately use the trowel to press a paper tape—used to reinforce the drywall and to hide imperfections—into the wet compound and to smooth away excess material. Nail and screw depressions also are covered with this compound, as are imperfections caused by the installation of air-conditioning vents and other fixtures. On large commercial projects, finishers may use automatic taping tools that apply the joint compound and tape in one step. Finishers apply second and third coats, sanding the treated areas after each coat to make them as smooth as the rest of the wall surface. This results in a very smooth and almost perfect surface. Some finishers apply textured surfaces to walls and ceilings with trowels, brushes, or spray guns.

Lathers apply metal or gypsum lath to walls, ceilings, or ornamental frameworks to form the support base for plaster coatings. Gypsum lath is similar to a drywall panel, but smaller. Metal lath is used where the plaster application will be exposed to weather or water, or for curved or irregular surfaces for which drywall is not a practical material. Lathers usually nail, screw, staple, or wire-tie the lath directly to the structural framework.

Plastering—one of the oldest crafts in the building trades— is enjoying a resurgence in popularity because of the introduction of newer, less costly materials and techniques. Plasterers apply plaster to interior walls and ceilings to form fire-resistant and relatively soundproof surfaces. They also apply plaster veneer over drywall to create smooth or textured abrasion-resistant finishes. They apply durable plasters such as polymer-based acrylic finishes and stucco to exterior surfaces, and install prefabricated exterior insulation systems over existing walls for good insulation and interesting architectural effects. In addition, they cast ornamental designs in plaster.

When plasterers work with interior surfaces such as cinder block and concrete, they first apply a brown coat of gypsum plaster that provides a base, followed by a second or finish coat—also called "white coat"—which is a lime-based plaster. When plastering metal lath (supportive wire mesh) foundations, they apply a preparatory or "scratch coat" with a trowel. They spread this rich plaster mixture into and over the metal lath. Before the plaster sets, they scratch its surface with a rake-like tool to produce ridges so the subsequent brown coat will bond to it tightly.

Laborers prepare a thick, smooth plaster for the brown coat. Plasterers spray or trowel this mixture onto the surface, then finish by smoothing it to an even, level surface.

For the finish coat, plasterers prepare a mixture of lime, plaster of paris, and water. They quickly apply this onto the brown coat using a "hawk"—a light, metal plate with a handle—a trowel, a brush, and water. This mixture, which sets very quickly, produces a very smooth, durable finish.

Plasterers also work with a plaster material that can be finished in a single coat. This thin-coat, or gypsum veneer plaster, is made of lime and plaster of paris and is mixed with water at the job site. It provides a smooth, durable, abrasion-resistant finish on interior masonry surfaces, special gypsum baseboard, or drywall prepared with a bonding agent.

Plasterers create decorative interior surfaces as well. They do this by pressing a brush or trowel firmly against the wet plaster surface and using a circular hand motion to create decorative swirls.

For exterior work, plasterers usually apply a mixture of portland cement, lime, and sand (stucco) over cement, concrete, masonry, and lath. Stucco is also applied directly to a wire lath with a scratch coat, followed by a brown coat and then a finish coat. Plasterers may also embed marble or gravel chips into the finish coat to achieve a pebblelike, decorative finish.

Increasingly, plasterers apply insulation to the exteriors of new and old buildings. They cover the outer wall with rigid foam insulation board and reinforcing mesh and then trowel on a polymer-based or polymer-modified base coat. They apply an additional coat of this material with a decorative finish.

Plasterers sometimes do complex decorative and ornamental work that requires special skill and creativity. For example, they mold intricate wall and ceiling designs. Following an architect's blueprint, they pour or spray a special plaster into a mold and allow it to set. Workers then remove the molded plaster and put it in place according to the plan.

As in other construction trades, drywall and lathing work sometimes is strenuous. Applicators, tapers, finishers, and lathers spend most of the day on their feet, either standing, bending, or kneeling. Some finishers use stilts to tape and finish ceiling and angle joints. Installers have to lift and maneuver heavy panels. Hazards include falls from ladders and scaffolds, and injuries from power tools. Because sanding joint compound to a smooth finish creates a great deal of dust, some finishers wear masks for protection.

Most plastering jobs are indoors; however, plasterers work outside when applying stucco or exterior wall insulation and decorative finish systems. Because plaster can freeze, heat is usually necessary to complete plastering jobs in cold weather. Sometimes plasterers work on scaffolds high above the ground.

Plastering is physically demanding, requiring considerable standing, bending, lifting, and reaching overhead. The work can be dusty and dirty; plaster materials also soil shoes and clothing and can irritate skin and eyes.

TRAINING

Most drywall, lathing workers, and plasterers start as helpers and learn their skills on the job. Others learn their trade in an apprenticeship program.

Installer and lather helpers start by carrying materials, lifting and holding panels, and cleaning up debris. Within a few weeks, they learn to measure, cut, and install materials. Eventually, they become fully experienced workers. Finisher apprentices begin by taping joints and touching up nail holes, scrapes, and other imperfections. They soon learn to install corner guards and to conceal openings around pipes. At the end of their training, they learn to estimate the cost of installing and finishing drywall and gypsum lath.

Plasterers who learn the trade informally as helpers usually start by carrying materials, setting up scaffolds, and mixing plaster. Later they learn to apply the scratch, brown, and finish coats.

The United Brotherhood of Carpenters and Joiners of America, in cooperation with local contractors, administers an apprenticeship program in carpentry that includes instruction in drywall and lath installation. In addition, local affiliates of the Associated Builders and Contractors and the National Association of Home Builders conduct training programs for nonunion workers. The International Brotherhood of Painters and Allied Trades conducts a two-year apprenticeship program for drywall finishers.

Apprenticeship programs, sponsored by local joint committees of contractors and unions, generally consist of two or three years of on-the-job training, in addition to at least 144 hours annually of classroom instruction in drafting, blueprint reading, and mathematics for layout work.

In the classroom, plasterer apprentices start with a history of the trade and the industry. They also learn about the uses of plaster, estimating materials and costs, and casting ornamental plaster designs. On the job, they learn about lath bases, plaster mixes, methods of plastering, blueprint reading, and safety. They also learn how to use various tools, such as hand and powered trowels, floats, brushes, straightedges, power tools, plaster-mixing machines, and piston-type pumps. Some apprenticeship programs also allow individuals to obtain training in related occupations such as cement masonry and bricklaying.

Applicants for apprentice or helper jobs generally must be at least 17 years old, be in good physical condition, and have manual dexterity. Applicants who have a high school education are preferred. Courses in general mathematics, mechanical drawing, and shop provide a useful background.

Drywall workers, lathers, and plasterers with a few years' experience and leadership ability may become supervisors. Some workers start their own contracting businesses.

JOB OUTLOOK

Drywall workers, lathers, and plasterers hold about 160,000 jobs nationwide. Most work for specialty contractors; others work for contractors who do many kinds of construction. Nearly one-third of drywall workers and lathers are self-employed, independent contractors. About one out of every five plasterers is self-employed.

Most installers, finishers, and lathers are employed in urban areas. In other areas, where there may not be enough work to keep a drywall worker or lather employed full time, the work is usually done by carpenters and painters.

Most plasterers work on new construction, particularly where special architectural and lighting effects are part of the work. Some repair and renovate older buildings.

Replacement needs will account for almost all job openings for drywall workers and lathers through the year 2005. Tens of thousands of jobs will open up each year because of the need to replace workers who transfer to jobs in other occupations or leave the labor force. Turnover in this occupation is very high, reflecting the lack of formal training requirements and the ups and downs of the business cycle, to which the construction industry is very sensitive. Because of their relatively weak attachment to the occupation, many workers with limited skills leave the occupation when they find they dislike the work or because they can't find steady employment.

Additional job openings will be created by the rising demand for drywall work. Employment is expected to grow more slowly than the average for all occupations, reflecting the slow growth of new construction and renovation. In addition to traditional interior work, the growing acceptance of insulated exterior wall systems will provide additional jobs for drywall workers.

Despite the growing use of exterior panels, most drywall installation, finishing, and lathing is usually done indoors. Therefore, these workers lose less work time because of bad weather than some other construction workers. Nevertheless, they may be unemployed between construction projects and during downturns in construction activity.

Employment of plasterers is expected to increase about as fast as the average for all occupations through the year 2005. In addition to job openings due to rising demand for plastering work, jobs will open up as plasterers transfer to other occupations or leave the labor force.

In past years, employment of plasterers declined as more builders switched to drywall construction. This decline has halted, however, and employment of plasterers is expected to continue growing as a result of greater appreciation for the durability and attractiveness that troweled finishes provide. Thin-coat plastering—or veneering—in particular is gaining greater acceptance as more builders recognize its ease of application, durability, quality of finish, and fire-retarding qualities. Prefabricated wall systems and new polymer-based or polymer-modified acrylic exterior insulating finishes are also gaining popularity, not only because of their durability, attractiveness, and insulating properties, but also because of their lower cost. These wall systems and finishes are growing in popularity particularly in the southern and southwestern regions of the country. In addition, plasterers will be needed to renovate plasterwork in older structures and create special architectural effects such as curved surfaces, which are not practical with drywall materials.

Most plasterers work in construction, where prospects fluctuate from year to year due to changing economic conditions. Bad weather affects plastering less than other construction trades because most work is indoors. On exterior surfacing jobs, however, plasterers may lose time because materials cannot be applied under wet or freezing conditions. Best employment opportunities should continue to be in Florida, California, and the Southwest, where exterior plaster and decorative finishes are expected to remain popular.

SALARIES

Median weekly earnings for drywall workers and lathers were about $430 in 1996. The middle 50 percent earned between $293 and $630 weekly. The top 10 percent earned more than $871, and the bottom 10 percent earned less than $204 a week. Trainees usually start at about half the rate paid to experienced workers and receive wage increases as they become more highly skilled.

Some contractors pay workers according to the number of panels they install or finish per day; others pay an hourly rate. A 40-hour week is standard, but sometimes the workweek may be longer. Those who are paid hourly rates receive premium pay for overtime.

Median weekly earnings for plasterers working full time were about $531 a week in 1996. The middle 50 percent earned between $354 and $760 a week. The top 10 percent earned more than $960, and the lowest 10 percent earned less than $251 a week.

According to the limited information available, average hourly earnings—including benefits—for plasterers who belonged to a union and worked full time ranged between $14.45 and $39.08 in 1997. Plasterers in New York, Boston, Chicago, San Francisco, Los Angeles, and other large cities received the higher hourly earnings. Apprentice wage rates start at about half the rate paid to experienced plasterers. Annual earnings for plasterers and apprentices may be less than the hourly rate would indicate because poor weather and periodic declines in construction activity may limit their work time.

Many plasterers are members of unions. They are represented by the Operative Plasterers' and Cement Masons' International Association of the United States and Canada, or the International Union of Bricklayers and Allied Craftsmen.

RELATED FIELDS

Drywall workers, lathers, and plasterers combine strength and dexterity with precision and accuracy to make materials fit according to a plan. Occupations that require similar abilities include carpenters, floor covering installers, form builders, and insulation workers.

INTERVIEW
Holly Perry
Drywall Hanger

Holly Perry works for Muddy Boys Drywall, commercial and residential contractors based in Utah. She has been in the field since 1992.

How Holly Perry Got Started

"My husband needed some help on a house he was working on, and so I offered. We then decided that since I did a good job for him and am very 'fussy,' we should pocket all the money ourselves, rather than pay it out to kids who wouldn't do as good of a job. We are very well liked and requested by contractors all the time. We do mostly custom homes.

"I learned everything by hands-on training. My husband taught me all the proper ways to measure, cut, and hang drywall. He taught me how to use the various tools of the trade, of which there are many—routers, screw guns, tape measurers, razors, and multiple hand tools.

"Once my husband and I completed our first house (and boy were we slow), the contractor threw us a few bones. Once we got our speed up to par, we kept the quality high. 'Women are much more meticulous and have more of an eye for quality,' the contractor said."

What the Job's Really Like

"When we enter a new job, it isn't a house yet. It's studs, insulation, tin, and outlets with wires hanging from them. Usually we have to clean up the leftovers from other subcontractors.

"We carry all our tools—boxes of nails/screws, aluminum benches and/or scaffolding, T squares, and extension cords—and pile them all on the floor.

"Then we do a walk-through to spot anything out of the norm—which could be anything from wires looped around a stud to missing backing. Spotting possible flaws that preceded us can save us a lot of time once we begin our job.

"We take an inventory of the Sheetrock to make sure that we have what the list says. Stockers can make mistakes, and they often break sheets and discard them.

"Next we do what I call an ounce of prevention. We walk through the house with a can of 'bright' red paint and spray on the floor the location of outlets, floor and ceiling vents, access holes, and pipes. Should something be missed and covered up with Sheetrock, when it comes time to clean up and finish 'screwing off' the house, it's nice to have the paint on the floor, telling us what is hidden behind the walls or ceiling.

"Now that we are ready to actually hang the drywall, we crank up the radio (preferably to an old rock-and-roll station) and hook up the power tools. We hang the lids (ceiling) first, then the top sheets, then the bottom sheets last.

"People always say that Sheetrock is so heavy. Sheetrock is heavy, but balance is the key. Picking up the sheet is easy with two people and a 'ready . . . set' count. Stepping up onto the benches with a sheet is tricky, and you have to have confidence in your footing and your partner's footing. Be slow and methodical.

"What I like least about hanging drywall is that so many times we run into inferior work done by framers, tinners, electricians, or plumbers. If a house isn't framed properly (which most are not), and studs aren't placed flush, it causes a ripple effect. Framers who leave nails protruding cause a lot of damage to Sheetrock, if we do not notice it prior to placing a sheet on the wall. And often because of overinsulation, nails are hidden from sight. Tinners who install heating and AC ducts/vent openings often don't secure them, and if they aren't secure, it can cause us to make sloppy cutouts. Many times we have had to tear off sheets and make repairs to tinners' work.

"Sheetrock is the finished product of a home. It is the one thing that shows every detail of all the work performed, and the one thing that makes home owners see their vision of what their new home will look like. It's the point where a home owner gets a real glint in the eye. And that is my favorite part of my job. Seeing a home owner's happiness makes it all worthwhile.

"I make about $15 an hour as an apprentice, but that fluctuates because we are mostly paid by the square foot. If the hanging-only portion of the house pays 15 cents a foot and the house is 12,000 square feet, then I would make one-half of $1,800, or $900, based on just two of us. A 12,000-square-foot house would normally take us about four to five days to hang, depending on the difficulty of the house plan (i.e., vaults, planter shelves, stairs, etc.). The money can be excellent if you have the right team working with you.

"A new person would be paid an hourly rate of $7 and, depending on how quickly he or she learns the ins and outs, could advance quickly up the pay scale. Newbies come and go a lot in this trade."

Expert Advice

"I could say that anyone can hang drywall, but it's not true. Anyone who wants to do it has to keep someone else's vision in mind. You have to follow basic rules, but even more, you have to see your work as a creation. Shoddy workmanship will follow you from job to job, but so will good work.

"Drywall is a 'hands-on' job. You can read 'how-to' books, but until you feel the weight and texture of gypsum, it's hard to get. I would suggest becoming familiar with the feel of the tools first. There is nothing like a big hammer or power tools. Getting to know the 'what you can and can't do' with Sheetrock is very important, too. Arched windows and entryways are Sheetrock, too. The trick is learning how to flex it; it just takes a certain knack.

"Many would think that I, as a woman in my field, was built like a bull, but I'm not. I am tall, and now I have muscles, but I am a very normal woman. A little Sheetrock dust (OK, a lot of Sheetrock dust) may not be pretty, but it sure feels good when I put my paycheck in the bank. The money is good. The personal satisfaction is more than rewarding. Pride in your work is all that you need to accomplish this goal, along with a good eye, and patience to see it through."

FOR MORE INFORMATION

For information about work opportunities in drywall application and finishing, contact local drywall installation contractors; a local of the unions mentioned; a local joint union-management apprenticeship committee; a state or local chapter of the Associated Builders and Contractors; or the nearest office of the state employment service or state apprenticeship agency.

For details about job qualifications and training programs in drywall application and finishing, write to:

Associated Builders and Contractors
1300 North 17th Street
Rosslyn, VA 22209

Home Builders Institute
National Association of Home Builders
1201 15th Street NW
Washington, D.C. 20005

International Brotherhood of Painters and Allied Trades
1750 New York Avenue NW
Washington, D.C. 20006

United Brotherhood of Carpenters and Joiners of America
101 Constitution Avenue NW
Washington, D.C. 20001

For information about plasterer apprenticeships or other work opportunities, contact local plastering contractors; locals of the unions mentioned; a local joint union-management apprenticeship committee; or the nearest office of the state apprenticeship agency or the state employment service.

For general information about the work of plasterers, contact:

International Union of Bricklayers and Allied Craftsmen
815 15th Street NW
Washington, D.C. 20005

Operative Plasterers' and Cement Masons' International
    Association of the United States and Canada
1125 17th Street NW
Washington, D.C. 20036

CHAPTER 11

Painters and Paperhangers

EDUCATION
H.S. Preferred

$$$ SALARY
$15,000 to $37,000

OVERVIEW

Paint and wall coverings make surfaces clean, attractive, and bright. In addition, paints and other sealers protect outside walls from wear caused by exposure to the weather. Although some people do both painting and paperhanging, each requires different skills.

Painters apply paint, stain, varnish, and other finishes to buildings and other structures. They choose the right paint or finish for the surface to be covered, taking into account customers' wishes, durability, ease of handling, and method of application. They first prepare the surfaces to be covered so the paint will adhere properly. This may require removing the old coat by stripping, sanding, wire brushing, burning, or water and abrasive blasting.

Painters also wash walls and trim to remove dirt and grease, fill nail holes and cracks, sandpaper rough spots, and brush off dust. On new surfaces, they apply a primer or sealer to prepare them for the finish coat. Painters also mix paints and match colors, relying on knowledge of paint composition and color harmony.

There are several ways to apply paint and similar coverings. Painters must be able to choose the right paint applicator for each job, depending on the surface to be covered, the characteristics of the finish, and other factors. Some jobs need only a good bristle brush with a soft, tapered edge; others require a dip or fountain pressure roller; still others can best be done using a paint sprayer. Many jobs need several types of applicators. The right tools for each job not only expedite the painter's work but also produce the most attractive surface.

When working on tall buildings, painters erect scaffolding, including "swing stages," scaffolds suspended by ropes or cables attached to roof hooks. When painting steeples and other conical structures, they use a "bosun chair," a swinglike device.

Paperhangers cover walls and ceilings with decorative wall coverings made of paper, vinyl, or fabric. They first prepare the surface to be covered by applying "sizing," which seals the surface and makes the covering stick better. When redecorating, they may first remove the old covering by soaking, steaming, or applying solvents. When necessary, they patch holes and take care of other imperfections before hanging the new wall covering.

After the surface has been prepared, paperhangers must prepare the paste or other adhesive. Then they measure the area to be covered, check the covering for flaws, cut the covering into strips of the proper size, and closely examine the pattern to match it when the strips are hung.

The next step is to brush or roll the adhesive onto the back of the covering, then to place the strips on the wall or ceiling, making sure the pattern is matched, the strips are hung straight, and the edges are butted together to make tight, closed seams. Finally, paperhangers smooth the strips to remove bubbles and wrinkles, trim the top and bottom with a razor knife, and wipe off any excess adhesive.

Most painters and paperhangers work 40 hours a week or less; about one out of six works part time. Painters and paperhangers must stand for long periods. Their jobs also require a considerable amount of climbing and bending. These workers must have stamina because much of the work is done with their arms raised overhead. Painters often work outdoors, but seldom in wet, cold, or inclement weather.

Painters and paperhangers risk injury from slips or falls off ladders and scaffolds. They may sometimes work with materials that can be hazardous if masks are not worn or if ventilation is poor.

TRAINING

Painting and paperhanging are learned through either apprenticeship or informal, on-the-job instruction. Although training authorities recommend completion of an apprenticeship as the best way to become a painter or paperhanger, most painters learn the trade informally on the job as a helper to an experienced painter. Few opportunities for informal training exist for paperhangers because few paperhangers have a need for helpers.

The apprenticeship for painters and paperhangers consists of three to four years of on-the-job training, in addition to 144 hours of related classroom instruction each year. Apprentices receive instruction in color harmony, use and care of tools and equipment, surface preparation, application techniques, paint mixing and matching, characteristics of different finishes, blueprint reading, wood finishing, and safety.

Whether a painter learns the trade through a formal apprenticeship or informally as a helper, on-the-job instruction covers similar skill areas. Under the direction of experienced workers, trainees carry supplies, erect scaffolds, and do simple painting and surface preparation tasks while they learn about paint and painting equipment. Within two or three years, trainees learn to prepare surfaces for painting and paper hanging, to mix paints, and to apply paint and wall coverings efficiently and neatly. Near the end of their training, they may learn decorating concepts, color coordination, and cost-estimating techniques.

Apprentices or helpers generally must be at least 16 years old and in good physical condition. A high school education or its equivalent that includes courses in mathematics is generally required to enter an apprenticeship program. Applicants should have manual dexterity and a good color sense.

Painters and paperhangers may advance to supervisory or estimating jobs with painting and decorating contractors. Many establish their own painting and decorating businesses.

JOB OUTLOOK

Painters and paperhangers hold about 449,000 jobs; most are painters. The majority of painters and paperhangers work for contractors engaged in new construction, repair, restoration, or remodeling work. In addition, organizations that own or manage large buildings, such as apartment complexes, employ maintenance painters, as do some schools, hospitals, and factories.

Remodeling, restoration, and maintenance projects often provide jobs for painters and paperhangers, even when new construction declines.

Self-employed independent painting contractors account for almost half of all painters and paperhangers, about twice the proportion of building trades workers in general.

Employment of painters and paperhangers is expected to grow about as fast as the average for all occupations through the year 2005 as the level of new construction increases and the stock of buildings and other structures that require maintenance and renovation grows. In addition to job openings created by rising demand for the services of these workers, many tens of thousands of jobs will become available each year as painters and paperhangers transfer to other occupations or leave the labor force. There are no strict training requirements for entry, so many people with limited skills work as painters or paperhangers for a short time and then move on to other types of work or work part time, creating many job openings. Many fewer openings will occur for paperhangers because the number of these jobs is comparatively small.

Prospects for people seeking jobs as painters or paperhangers should be quite favorable, because of the high turnover.

Despite the favorable overall conditions, job seekers considering these occupations should expect some periods of unemployment because many construction projects are of short duration and construction activity is cyclical and seasonal in nature.

Remodeling, restoration, and maintenance projects, however, often provide many jobs for painters and paperhangers even when new construction activity declines. The most versatile painters and paperhangers generally are most able to keep working steadily during downturns in the economy.

SALARIES

Median weekly earnings for painters who are not self-employed are about $391. Most earn between $288 and $516 weekly. The top 10 percent earn more than $721.

In general, paperhangers earn more than painters. Earnings for painters may be reduced on occasion because of bad weather and the short-term nature of many construction jobs.

Hourly wage rates for apprentices usually start at 40 to 50 percent of the rate for experienced workers and increase periodically.

Some painters and paperhangers are members of the International Brotherhood of Painters and Allied Trades. Some maintenance painters are members of other unions.

RELATED FIELDS

Painters and paperhangers apply various coverings to decorate and protect wood, drywall, metal, and other surfaces. Other occupations in which workers apply paints and similar finishes include billboard posterers, metal sprayers, undercoaters, and transportation equipment painters.

INTERVIEW
Dan Brawner
Painter

Dan Brawner is the owner of Brawner Painting in Lisbon, Iowa. He began working for others in 1983 and set out on his own seven years ago.

How Dan Brawner Got Started

"My father was a house painter and general contractor. I started working for him after school and in the summers when I was about 8 years old. I was a spray painter and a regular member of the crew by the time I was 10. When I was 13, my older brother and I went on the road, seven days a week, living in a tent, painting farms across Iowa.

"As a teenager, I thought it was hard work and messy and there was nothing cool about it. Then, after many lean years as a freelance writer, it dawned on me that there are a couple of things that are very attractive about painting. First was the money. Painting was something I grew up with. I knew the business, and I could do the work in my sleep. I really had no trouble starting up the business and getting customers. Second was the fact that in Iowa, exterior painting is seasonal work. If I worked like a maniac (what other way is there?) for five months, I could write for seven months. Dr. Samuel Johnson (18th-century writer) said nobody but a fool would write for any reason but money. Painting allows me to indulge my foolishness.

"My first painting job was, of course, with my father. After getting an elite education that made me too good for a lowly trade like painting, I set my sights higher.

"Years later and a few pounds lighter, I started my own painting business. I may grumble that house painting is the last refuge of the unemployable, but to paraphrase Fernando Valenzuela, painting has been veddy, veddy good to me."

What the Job's Really Like

"As the owner/operator of the painting business, I do everything. Most important, I do it when and how I promised I would. Small contractors have a reputation for being flaky, so I make an effort to show up as expected and bring in the project on time and under budget. Public relations is a big part of my job. The work isn't done until the customer is happy.

"Because my job is seasonal, I work as long as there is light in the sky. Things such as cleaning brushes and machinery can be done after dark. I like what I do because of the variety of tasks—the advertising, bidding, and painting, for example.

"Mostly, I like the details. OK, I *love* the details. I love going back after doing a respectable job and adding those touches that make it a remarkable job. That's not bragging. I think every contractor does this. After all the equipment breakdowns and bad weather and sore muscles, taking the time for the finishing touches gives me a reason to get up in the morning.

"What I like most about the painting business is standing back and admiring the finished product. I paint Victorian houses with seven or more colors. When the builders who made these things nailed the last board in place, they must have given each one a special curse, designed to kill any contractor who dared to touch their precious house from then on. Victorians have the oddest angles and strangest, broken roof lines. It had to be intentional, the way the windows were spaced so no ladder could be set between them. The fish scale shingles and dental trim have deep crevices that ruin brushes and drink up your paint and your time. But when one of these old monsters is freshly painted, it is a thing of beauty.

"I started out painting farm buildings with red paint and white trim. On a shimmering August afternoon in Iowa, with the cornfields in the background, punctuated by lush, round, Grant Wood trees, the simple elegance of a red barn with white trim still takes my breath away.

"Beginning painters often start out at $7 or $8 an hour. Spray painters, powder painters, body shop painters, and other specialists can expect to make $20 to $40 an hour or more. Since I started my painting business, I have literally earned as much as I wanted. That's not to say it's a lot by most people's standards. But the beauty of the business is that when I need more, all I have to do to get it is work more. It sounds simple enough, but this cause-and-effect relationship between effort and reward was not something I ever saw working for somebody else."

Expert Advice

"Beginning painters should have a critical eye for detail and a tolerance for repetitive movement. Being in good physical condition helps, and it doesn't hurt to be extremely tall. (Of course, you could use ladders.) Resilience is a good quality for painters. Last summer, while painting a low overhang, I walked off the edge of a flat roof and hit hard. But fortunately, I bounced right back.

"If you want a job as a painter, you could start out painting your own house or apartment. You could practice (for money) on the homes of friends and relatives. Then, if you haven't got a regular painting job by September, all you have to do is walk out to a job site in a pair of white overalls and a white T-shirt and tell the foreman you're looking for work. And be sure you're wearing old shoes, because while the foreman is expressing his gratitude, he's likely to slobber all over them."

INTERVIEW
Michael Keith
Paperhanger

Michael Keith is a self-employed paperhanger in San Jose, California. He has been in the field since 1971 and has won first-place awards in national trade competitions for residential and commercial wall covering. He was also the second vice president of the National Guild of Professional Paperhangers from 1994 through 1997.

How Michael Keith Got Started

"I became a painter first, an allied field, because the work seemed interesting, not too aerobic, and job satisfaction seemed immediate, though at age 18 I don't remember job satisfaction as being that important a consideration.

"I began to install wallpaper because it was a closely allied occupation with increasing demand, and again the satisfaction was instantaneous.

"I worked for a painting contractor who also installed wall covering. When he hung paper, I assisted, learning the theory and technique.

"In 1971 I was working for myself, painting a house. The client, a real estate agent who had a severe drinking problem, mentioned he'd like to paper a wall and wondered if I did that. I said, sure, thinking that if I screwed it up, his constant intoxication would preclude his noticing the difference. I wallpapered the wall successfully and found it to be enjoyable, so I added the word *Paperhanger* to my business card. A year later, I removed the word *Painter*."

What the Job's Really Like

"My job is to install a wide variety of wall coverings, requiring a fairly extensive knowledge of product and procedure. Additional knowledge of painting, drywall, and plastering procedures is also necessary.

"Installations usually require the preparation, smoothing, and proper priming of wall surfaces to make them ready for wall coverings. Often, though, the preparation has been completed prior to my arrival, and then my task is simply to install the wall covering.

"A typical day involves arriving at the installation site and setting up my equipment, which consists of a portable table for the trimming and pasting of the wall covering, a small machine used to apply the adhesive to the wall covering, a selection of hand tools used for the process, containers of adhesive, a bucket and sponges for rinse water, and ladders or scaffolding as needed for that particular application.

"The next step is to engineer the room. Experience, the dimension of the wallpaper and its ratio to the dimensions of the room, and the architecture of the space will determine the starting point.

"Adhesive is applied to the back of the wall covering, which is then cut to size, book folded, paste side to paste side, and put aside for a few minutes to marinate—this allows the material to absorb moisture and expand. The pasted wall covering is then taken to the wall, unfolded, and applied, using a plumb guideline. A second strip is applied to the wall, carefully placed in relationship to the first strip, using the pattern as a reference. Constant adjustments are required to maintain the pattern and to take into account any architectural variations. This process is repeated until all the intended walls are covered with the wall covering.

"Depending on the type of material and the surrounding working conditions, the work atmosphere can range from relaxed to frantic. The presence of other trades and the accompanying noise and intrusions can affect conditions greatly.

"There is a marked difference between commercial and residential installation in this respect. With residential work, the paperhanger is usually the only worker on the site, while with commercial work, there are usually a number of trades present, and territory conflicts can occur.

"The trade-off is that commercial work will usually involve the installation of industrial vinyl, which is inherently less problematic than many residential goods.

"Though the work may seem from appearances to be repetitive, it is never boring. Differences in materials and working conditions as well as architectural differences and anomalies keep the work interesting. There is a simplistic, Zenlike quality to the work. In my workday there is only me and the wall, and then suddenly, I'm done.

"I never watch the clock, and when I look at one, it is always later than I think. Time flies. (This attitude toward the clock is typical of self-employed people but not so for people who are employed by others.) I am often alone at the work site, and there is a quality of solitude that I very much enjoy.

"My workweek is usually 30 to 40 hours except in times of hurry, when an occasional weekend is absorbed, but this is by choice, usually to accommodate the business schedule of commercial or professional offices and the like, and occasionally to accommodate my own schedule, if I'm running behind.

"When you have observers around while you are working, they are often very interested in the process. Many have tried to hang paper and have a respect for and awe of the craft.

"What I like least about the work is shoddy, substandard materials that continue to appear on the market, and the thoroughly unpleasant people that one encounters from time to time in most any venue.

"But after 27 years, I still find the work satisfying. Do I like what I do? I love what I do. What I like most about my work is that I can enter a residence or place of business and within a day or a week, I drastically change the look of the space, much to the satisfaction and awe of the client. Job satisfaction is immediate, and everything I get on me comes off in the shower. Though some people have tried to convince me to the contrary, wallpaper is never a life-or-death matter."

Expert Advice

"I have had conversations with rich and successful business people who have expressed an envy for my working conditions and lifestyle, and the fact that I work with my hands and can see the immediate results of that work.

"People who lean toward left-brain tendencies (or is that right brain?—I get 'em mixed up) and working with their hands on things that are real and concrete, and are not excited about abstract tasks where job completion and satisfaction is conceptual and subject to interpretation, may very well be candidates for this work.

"I'm told that paperhanging requires patience. I don't view myself as particularly patient, and I certainly have thrown my share of wrenches, but I suppose that the trait that causes one to stick to something until it is right probably is patience. If you're willing to spend the time required to complete a task correctly, this craft may be for you.

"If you do wish to pursue this craft, there is much to be done before printing business cards. Unless you are willing to learn at the expense of the client—not the best of ideas—you must acquire the trade theory and procedure before offering your service without supervision.

"Though there is not a glut of wallpaper schools in this country, there are a few here and there. The American School of Paperhanging Arts in Commerce, Georgia, though no longer holding regular school sessions, does offer a comprehensive video course and technical support that is respected in the industry.

"There are also formal apprenticeships available through local trade unions that encompass the craft.

"Another way to acquire the basics and fine-tune the particulars is to align yourself with an established professional and seek employment or a working affiliation. These people have already made all the possible mistakes, and you can learn vast amounts from these veterans.

"Also, on-line information for the paperhanger is becoming increasingly available. There are currently two separate networks of paperhangers on the Internet and several others based on the painting trade that include paperhanging information. All of this is available with a computer and a modem.

"This is a craft for men and women, and my personal observation is that women inherently tend toward the precision and meticulousness that prove to be a real plus in this type of endeavor.

"Additionally, in residential work, the client is usually a woman and will tend to be more comfortable when the craftsperson in her home is a woman.

"I have also noticed that many successful paperhangers evolve into a specialty, tending toward a niche market that they find comfortable, and one at which they are particularly good. You may find that you really like textiles or high-end hand screens. I know a paperhanger who does only historic restoration. He recently installed a custom wallpaper in the Blue Room in the White House.

"As far as salaries go, I live and work as a self-employed paperhanger in the Silicon Valley of California where I can expect to earn from $40 to $60 per hour, depending on clientele and degree of difficulty. It should be noted that the cost of living here is considerably higher than in other regions of the country. Wage rates vary as you move across the map. The level and range of your experience and your marketing ability, combined with location and local economics, will determine how much money you can make.

"A different wage schedule is involved for someone who is employed by a wall covering contractor. You should decide, when pursuing a craft, if you want to participate only at the craft level, as an employee, or if you want to assume the responsibility of contracting the work, which involves myriad other skills and activities.

"I have noticed that running a business attracts a different breed of person from one who is simply pursuing a craft. It is not for everyone."

FOR MORE INFORMATION

For details about painting and paperhanging apprenticeships or work opportunities, contact local painting and decorating contractors; a local of the International Brotherhood of Painters and Allied Trades; a local joint union-management apprenticeship committee; or an office of the state apprenticeship agency or state employment service.

For additional information about the work of painters and paperhangers, contact:

American School of Paperhanging Arts
Little and Oak Streets
Commerce, GA 30529

Associated Builders and Contractors
1300 North 17th Street
Rosslyn, VA 22209

International Brotherhood of Painters and Allied Trades
1750 New York Avenue NW
Washington, D.C. 20006

Home Builders Institute
National Association of Home Builders
1201 15th Street NW
Washington, D.C. 20005

National Guild of Professional Paperhangers
910 Charles Street
Fredricksburg, VA 22401

CHAPTER 12 · Construction and Building Inspectors

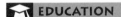

EDUCATION
B.A./B.S. Preferred

$$$ SALARY
$21,000 to $56,000

OVERVIEW

Construction and building inspectors examine the construction, alteration, or repair of buildings, highways and streets, sewer and water systems, dams, bridges, and other structures to ensure compliance with building codes and ordinances, zoning regulations, and contract specifications. Inspectors generally specialize in one particular type of construction work or construction trade, such as electrical work or plumbing. They make an initial inspection during the first phase of construction, and follow-up inspections throughout the construction project to monitor compliance with regulations. In areas where severe natural disasters—such as earthquakes or hurricanes—are more common, inspectors monitor compliance with additional safety regulations.

Building inspectors inspect the structural quality and general safety of buildings. Some specialize—for example, in structural steel or reinforced concrete structures. Before construction begins, plan examiners determine whether the plans for the building or other structure comply with building code regulations and are suited to the engineering and environmental demands of the building site. Inspectors visit the work site

135

before the foundation is poured to inspect the soil condition and positioning and depth of the footings. Later they return to the site to inspect the foundation after it has been completed. The size and type of structure and the rate of completion determine the number of other site visits they must make. Upon completion of the entire project, they make a final comprehensive inspection.

A primary concern of building inspectors is fire safety. They inspect the structure's fire sprinklers, alarms, and smoke control systems, as well as fire doors and exits. In addition, inspectors may calculate fire insurance rates by assessing the type of construction, building contents, adequacy of fire protection equipment, and risks posed by adjoining buildings.

Electrical inspectors inspect the installation of electrical systems and equipment to ensure that they function properly and comply with electrical codes and standards. They visit work sites to inspect new and existing wiring, lighting, sound and security systems, motors, and generating equipment. They also inspect the installation of the electrical wiring for heating and air-conditioning systems, appliances, and other components.

Elevator inspectors examine lifting and conveying devices such as elevators, escalators, moving sidewalks, lifts and hoists, inclined railways, ski lifts, and amusement rides.

Mechanical inspectors inspect the installation of the mechanical components of commercial kitchen appliances, heating and air-conditioning equipment, gasoline and butane tanks, gas and oil piping, and gas-fired and oil-fired appliances. Some specialize in inspecting boilers or ventilating equipment.

Plumbing inspectors examine plumbing systems, including private disposal systems, water supply and distribution systems, plumbing fixtures and traps, and drain, waste, and vent lines.

Public works inspectors ensure that federal, state, and local government construction of water and sewer systems, highways, streets, bridges, and dams conforms to detailed contract specifications. They inspect excavation and fill operations, the placement of forms for concrete, concrete mixing and pouring, asphalt paving, and grading operations. They record the work and materials used so that contract payments can be calculated.

Public works inspectors may specialize in highways, structural steel, reinforced concrete, or ditches. Others specialize in dredging operations required for bridges and dams or for harbors.

Home inspectors conduct inspections of newly built homes to check that they meet all regulatory requirements. Home inspectors are also increasingly hired by prospective home buyers to inspect and report on the condition of a home's major systems, components, and structure. Typically, home inspectors are hired either immediately prior to a purchase offer on a home or as a contingency to a sales contract.

Construction and building inspectors increasingly use computers to help them monitor the status of construction inspection activities and to keep track of permits that were issued. Details about construction projects, building and occupancy permits, and other documentation are now generally stored on computers so that they can easily be retrieved and kept accurate and up-to-date.

Although inspections are primarily visual, inspectors often use tape measures, survey instruments, metering devices, and test equipment such as concrete strength measurers. They keep a daily log of their work, take photographs, file reports, and, if necessary, act on their findings. For example, construction inspectors notify the construction contractor, superintendent, or supervisor when they discover a code or ordinance violation or something that does not comply with the contract specifications or approved plans. If the problem is not corrected within a reasonable or specified period of time, government inspectors have authority to issue a "stop-work" order.

Many inspectors also investigate construction or alterations being done without proper permits. Violators of permit laws are directed to obtain permits and submit to inspection.

Nearly 60 percent of all construction and building inspectors work for local governments.

Construction and building inspectors usually work alone. However, several may be assigned to large, complex projects, particularly because inspectors specialize in different areas of construction. Though they spend considerable time inspecting construction work sites, inspectors may spend much of their days in a field office reviewing blueprints, answering letters or telephone calls, writing reports, and scheduling inspections.

Inspection sites are dirty and may be cluttered with tools, materials, or debris. Inspectors may have to climb ladders or many flights of stairs, or may have to crawl around in tight spaces. Although their work is not considered hazardous, inspectors usually wear "hard hats" for safety.

Inspectors normally work regular hours. However, if an accident occurs at a construction site, inspectors must respond immediately and may work additional hours to complete the report.

TRAINING

Individuals who want to become construction and building inspectors should have a thorough knowledge of construction materials and practices in either a general area, such as structural or heavy construction, or a specialized area, such as electrical or plumbing systems, reinforced concrete, or structural steel. Construction or building inspectors need several years of experience as a manager, supervisor, or craft worker before becoming inspectors. Many inspectors have previously worked as carpenters, electricians, plumbers, or pipe fitters.

Employers prefer to hire inspectors who have formal training as well as experience. Employers look for people who have studied engineering or architecture, or who have a degree from a community or junior college, with courses in construction technology, blueprint reading, mathematics, and building inspection. Courses in drafting, algebra, geometry, and English are also useful. Most employers require inspectors to have a high school diploma or equivalent, even when they qualify on the basis of experience.

Certification can enhance an inspector's opportunities for employment and advancement to more responsible positions. Most states and cities require some type of certification for employment. To become certified, inspectors with substantial experience and education must pass stringent examinations on code requirements, construction techniques, and materials.

Many categories of certification are awarded for inspectors and plan examiners in a variety of disciplines, including the designation "CBO," Certified Building Official. (Organizations that administer certification programs are listed at the end of this chapter.)

Construction and building inspectors must be in good physical condition in order to walk and climb about construction sites. They also must have a driver's license. In addition, federal, state, and many local governments may require that inspectors pass a civil service examination.

Construction and building inspectors usually receive most of their training on the job. At first, working with an experienced inspector, they learn about inspection techniques; codes, ordinances, and regulations; contract specifications; and record-keeping and reporting duties. They usually begin by inspecting less complex types of construction, such as residential buildings, and then progress to more difficult assignments. An engineering or architectural degree is often required for advancement to supervisory positions.

Because they advise builders and the general public on building codes, construction practices, and technical developments, construction and building inspectors must keep abreast of changes in these areas. Many employers provide formal training programs to broaden inspectors' knowledge of construction materials, practices, and techniques. Inspectors who work for small agencies or firms that do not conduct training programs can expand their knowledge and upgrade their skills by attending state-sponsored training programs, by taking college or correspondence courses, or by attending seminars sponsored by the organizations that certify inspectors.

JOB OUTLOOK

Construction and building inspectors hold about 64,000 jobs nationwide. More than 50 percent work for local governments, primarily municipal or county building departments. Employment of local government inspectors is concentrated in cities and in suburban areas undergoing rapid growth. Local governments employ large inspection staffs, including many plan examiners or inspectors who specialize in structural steel, reinforced concrete, boiler, electrical, and elevator inspection.

About 18 percent of all construction and building inspectors work for engineering and architectural services firms, conducting inspections for a fee or on a contract basis. Most of the remaining inspectors are employed by the federal and state governments.

Many construction inspectors employed by the federal government work for the U.S. Army Corps of Engineers or the General Services Administration. Other federal employers include the Tennessee Valley Authority and the Departments of Agriculture, Housing and Urban Development, and Interior.

Employment of construction and building inspectors is expected to grow faster than the average for all occupations through the year 2005. Increased concern for public safety and improvements in the quality of construction should continue to stimulate demand for construction and building inspectors. Despite the expected employment growth, most job openings will arise from the need to replace inspectors who transfer to other occupations or who leave the labor force. Replacement needs are relatively high because construction and building inspectors tend to be older, more experienced workers who have spent years working in other occupations.

Opportunities to become a construction and building inspector should be best for highly experienced supervisors and craft workers who have some college education, who have some engineering or architectural training, or who are certified as inspectors or plan examiners. Thorough knowledge of construction practices and skills in areas such as reading and evaluating blueprints and plans is essential.

Governments—particularly federal and state—should continue to contract out inspection work to engineering, architectural, and management services firms as their budgets remain tight. However, the volume of real estate transactions will increase as the population grows, and greater emphasis on home inspections should result in rapid growth in employment of home inspectors.

Inspectors are involved in all phases of construction, including maintenance and repair work, and are therefore less likely to lose jobs during recessionary periods when new construction slows.

SALARIES

The median annual salary of construction and building inspectors was around $33,700 in 1996. The middle 50 percent earned between $26,500 and $45,800. The lowest 10 percent earned less than $21,600, and the highest 10 percent earned more than $55,800 a year.

Generally, building inspectors, including plan examiners, earn the highest salaries. Salaries in large metropolitan areas are substantially higher than those in small local jurisdictions.

RELATED FIELDS

Construction and building inspectors combine a knowledge of construction principles and law with an ability to coordinate data, diagnose problems, and communicate with people. Workers in other occupations using a similar combination of skills include drafters, estimators, industrial engineering technicians, surveyors, architects, and construction contractors and managers.

INTERVIEW
Darrell Phelps
Construction Inspector

Darrell Phelps is employed by the Board of County Commissioners, Osceola County Public Works, in Kissimmee, Florida. He has been working in this field since 1985.

How Darrell Phelps Got Started

"I chose this field, in part, because of my background. As I was growing up, my father, a farmer, often took work on highway projects when crop and weather constraints allowed. He started by using the three mules he had trained to work roads and graduated to operating bulldozers and motor graders. It was always a thrill for us boys to be allowed up on the equipment while he was working. Being a liberal child-rearer, he even took my two sisters for a ride occasionally.

"After retiring from the Air Force in 1978, having training in explosives and construction, courtesy of the Air Force, I went looking for work. The best money and opportunity for advancement offered around Austin, Texas, where I was at the time, was in the construction field.

"Minimum wage then was $3.25, and I could start at $5, quickly advancing to $6 in less than six months. By learning how to operate all the equipment, working through lunch and after hours, I never ran out of work. By 1985, I was making $10 an hour and was provided with a vehicle to drive. That year, I was offered a job with the city of Austin as a blasting inspector and public works inspector. The city offered a 160-hour course (all done during work hours) from Texas A&M University. I took the course, which covered all aspects of construction: concrete placement, pipe laying, trench safety, OSHA rules, paving, drainage, and so forth.

"Wanting a change, I moved to Florida. I worked for a construction firm for six months, then applied for this job with Osceola County. With my construction background, plus being trained by the Air Force and the city of Austin, I was hired."

What the Job's Really Like

"I go to the office at 7 A.M., receive my daily assignments of driveway inspections and drainage complaints, then proceed to my area of responsibility. Any construction involving streets and drainage within my assigned area is under my jurisdiction. This includes storm drain, roadway subgrade, base material, paving, curbing, sidewalks, driveways and any other work within the right-of-way, and utility lines of any kind.

"I inspect construction, utilizing approved plans, Florida Department of Transportation specifications, and the county roadway specifications. Any house being built must conform to the drainage code in the grading of the lot.

"Some days are relaxed, and some are hectic. When I can schedule my day, it is fairly relaxed, but being a public servant, I must respond to citizens' complaints. Working for elected officials, I must always keep in mind 'Customer Service.' This can screw up a schedule.

"What I like the most is the outdoors. The job has a minimal amount of paperwork; all work is defined by plans and specifications. And I get to work with professionals most of the time.

"What I like the least are rainy days, citizens who expect something for nothing or more than is possible, and unprofessional contractors who cut corners."

Expert Advice

"When you're just starting out, work in related fields; in other words, get a job doing construction. Also get educated in the field. Most colleges have trade schools, and some universities such as Texas A&M have classes.

"You must have the ability to relate to people in a positive manner. You must be able to determine the possibilities within your parameters and to convince the complainants to accept the inevitable."

INTERVIEW
Dick Chance
Owner of Chance Inspection Services, Inc.

Dick Chance has been in the field since 1962. In 1965 he earned his bachelor's degree in civil engineering from Bradley University in Peoria, Illinois. In 1998 he started his own company, Chance Inspection Services, Inc., located in Pineville, Louisiana.

How Dick Chance Got Started

"I always wanted to be involved in construction in one way or another. I began as a technician for the State of Illinois Department of Transportation and progressed into the consulting engineering business.

"I had also worked for the American Red Cross as chief of damage assessment/building and repair for a 17-state area. The Red Cross relocated me twice, and the idea of a third move in nine years (relocation to the D.C. area) made me decline the last relocation. I decided to open my own business in inspection of residential and commercial buildings and providing construction management services.

"I expect to have a salary of approximately $38,000 during my first year in the inspection business. Start-up costs have run about $5,000. During my first month in business, I made contacts with all of the major real estate firms, developers, contractors, and the general public. I used the local newspaper in announcing the opening of my business and did minimal advertising. I have developed marketing brochures and spoken to the sales staff at most realty offices. Word-of-mouth advertising by satisfied clients has outdone paid advertising about 80 to 1.

"My fee schedule is based on square footage in residential inspections and an hourly rate for commercial and construction management. For example, a complete home inspection up to 2,400 square feet runs $200. Commercial and construction management rates start at $35 per hour, depending on the complexity of services needed. I am projecting an income of $50,000 for my second year in business."

What the Job's Really Like

"My job is to provide complete inspections of everything from single-family homes, apartment complexes, and commercial buildings to construction management of projects involving new construction.

"My job is very interesting and in a relaxed setting. I set my own schedule, do my own marketing, provide a needed service to buyers of real estate, provide easy-to-read written reports, help developers in the management of construction projects as the owners' representative, and use my experience to answer questions and give my clients peace of mind when they are making a large investment.

"A typical day begins early and ends late. It may include doing a complete home inspection in the morning, visiting construction sites to monitor progress and quality control in the afternoon, writing reports in the evening, and keeping up the business books and continually updating clients with telephone reports. Marketing continues at all times, and business networking is ongoing.

"Being in the inspection business never gets boring. No two inspections are alike, and some construction problems never cease to amaze me. I usually spend about 50 to 60 hours per week in the operation of my business. Some days are 'downright dirty' from crawling under structures and going through hot attics. Others are very enjoyable, and new construction inspections are supreme.

"I inspect all components that are visible in a bottom-to-top inspection, including the structure, roof, chimneys, fireplaces, plumbing, heating, air-conditioning, electrical, appliances, swimming pools, and spas/hot tubs. Inspection also includes the site, drives, walks, patios, and decks. I leave the possible insect infestations up to the 'bug man.'

"The best part of my job is working with all types of people and providing a good service and product. I may be working with a first-time buyer of a home or with an experienced contractor or developer. Listening is probably the most important aspect of the inspection business, and not jumping to conclusions but instead doing a thorough investigation of a problem and researching materials to see if anyone has documented this type of problem and how it was resolved. The use of computers and the Internet has really helped in the research process.

"The biggest drawback to this type of business is the local economy and the up-and-down interest rates. Being in the South makes it a year-round business, with the heat in the summer being almost unbearable when I'm inspecting attic areas. Outside temperatures may be 100 degrees or higher, with temps in attics reaching 130 or higher.

"Another downside is having to inspect crawl spaces with spiders, snakes, and other creatures lurking in dark places. You have to be very observant of the conditions around you at all times."

Expert Advice

"Someone entering the inspection business should have a well-rounded background in all aspects of construction. This person has to be very ethical and not waver in professionalism. If you provide good service to go along with a good product, you should succeed.

"Training is available through trade schools, correspondence schools, independent training seminars, and on-the-job training throughout the country. Several professional organizations also provide training and programs for business development. The American Society of Home Inspectors is one of the better ones, and they have much information for those wanting to enter the inspection business. They also have a code of conduct and ethics, leading the industry."

FOR MORE INFORMATION

Information about a career and certification as a construction or building inspector is available from the following model code organizations:

Building Officials and Code Administrators International, Inc.
4051 West Flossmoor Road
Country Club Hills, IL 60478

International Conference of Building Officials
5360 Workman Mill Road
Whittier, CA 90601-2298

Southern Building Code Congress International, Inc.
900 Montclair Road
Birmingham, AL 35213

Information about a career as a home inspector is available from:

American Society of Home Inspectors, Inc.
85 West Algonquin Road
Arlington Heights, IL 60005

For information about a career as a state or local government construction or building inspector, contact your state or local employment service.

About the Author

A full-time writer of career books, Blythe Camenson stresses that her main concern is helping job seekers make educated choices. She firmly believes that with enough information, readers can find long-term, satisfying careers. To that end, she researches traditional as well as unusual occupations, talking to a variety of professionals about what their jobs are really like. In all of her books she includes firsthand accounts from people who can reveal what to expect in each occupation, the upsides as well as the downsides.

Camenson's interests range from history and photography to writing novels. She is also director of Fiction Writer's Connection, a membership organization providing support to new and published writers.

Camenson was educated in Boston, earning her B.A. in English and psychology from the University of Massachusetts and her M.Ed. in counseling from Northeastern University.

In addition to *On the Job: Real People Working in Building and Construction* Blythe Camenson has written more than 30 books for NTC/Contemporary Publishing Group.